Management and Engineering Innovation

Management and Engineering Innovation

Edited by
Carolina Machado
J. Paulo Davim

iSTE

WILEY

First published 2013 in Great Britain and the United States by ISTE Ltd and John Wiley & Sons, Inc.

ISTE Ltd
27-37 St George's Road
London SW19 4EU
UK

www.iste.co.uk

John Wiley & Sons, Inc.
111 River Street
Hoboken, NJ 07030
USA

www.wiley.com

Library of Congress Control Number: 2013934675

British Library Cataloguing-in-Publication Data
A CIP record for this book is available from the British Library
ISBN: 978-1-84821-554-2

Table of Contents

Preface

This book entitled *Management and Engineering Innovation* puts a special emphasis on human resource management (HRM) and innovation. It provides discussion and an exchange of information on principles, strategies, models, techniques, methodologies and applications of management innovation in the field of industry, commerce and services. It aims to communicate the latest developments and thinking on the subject of management worldwide. This book links management and engineering disciplines, in an innovation context, to promote sustainable development. It seeks cultural and geographic diversity in studies of HRM and engineering and uses those that have a special impact on organizational communications, change processes and work practices, reflecting the diversity of societal and infrastructural conditions.

The book mainly analyzes research on important issues that transcend the boundaries of single academic subjects and managerial functions. It takes into account interdisciplinary scholarship and commentaries that challenge the paradigms and assumptions of individual disciplines or functions, which are based on conceptual and/or empirical literature. This book is designed to increase the knowledge and effectiveness of all those involved in management and engineering innovation whether in the profit or not-for-profit sectors, or the public or private sectors.

It covers management and engineering innovation in 8 chapters. Chapter 1 discusses engineers and managers, and their relationship using some case studies. Chapter 2 covers strategic capabilities for successful engagement in proactive CSR in small and medium enterprises from a resource-based view. Chapter 3 contains information on innovative management development in the automotive supply industry presenting a preliminary case study for the development of an innovative approach to innovation management. Chapter 4 describes innovative product design and development through online customization. Subsequently, Chapter 5 covers struggling for survival and success and whether Brazil's defense industry can help foster innovation. Chapter 6 contains information on knowledge management practices fostering innovation through an integrated approach. Chapter 7 describes institutional logics promoting and inhibiting innovation. Finally, Chapter 8 presents HRM in SMEs in Portugal by presenting an innovative proposal of characterization.

This book can be used as a textbook for final undergraduate engineering and/or management courses or as a subject on management and engineering innovation at the postgraduate level. It can also serve as a useful reference for academics, researchers, managers, engineers and other professionals in areas related to management and engineering innovation. The importance of this book is evident for many institutes and universities throughout the world.

The editors acknowledge their gratitude to ISTE Ltd. and John Wiley and Sons for providing this opportunity and for their professional support. Finally, they would like to thank all the chapter authors for their interest and availability to work on this project.

Carolina Feliciana MACHADO
Braga, PORTUGAL

J. Paulo DAVIM
Aveiro, PORTUGAL

Chapter 1

We the Engineers and Them the Managers

1.1. Introduction

People have managed since the beginning of time. When men were hunter gatherers, women managed families. The development of agriculture involved developing new management skills and techniques [MIT 99]. The Mandarins who administered imperial China were professional managers, and had training in mathematics, sciences and engineering [NEE 86]. Engineer-managers were crucial to the success of the Roman Empire, building roads that connected it and aqueducts that enabled urbanization [CHA 08, HOD 01]. In the Middle Ages, masons were both engineers and managers with their own highly effective and multinational professional association [BLO 68], and guilds that insisted on training and professional recognition for membership [DUR 57] were also central to the emergence of civic institutions [BRA 82]. Brunelleschi was an engineer-manager, inventing new engineering techniques and directly supervising the construction of the dome for the *Duomo* in Florence [COO 90]. Napoleon established the *Grandes Écoles* for engineer-managers that still flourish in modern France. From the early to mid 19th Century, Isambard Kingdom Brunel, who built the first iron ship, bridges, docks and railways, was a mechanical and civil engineer who directly managed their construction [VAU 91].

Chapter written by Teresa Carla TRIGO OLIVEIRA and João FONTES DA COSTA.

With a transition from entrepreneurial to modern capitalism [BER 32, BUR 62], this changed. At a technical level, engineers became increasingly important. Yet when engineering breakthroughs such as in electricity and telephony emerged, and were applied by those such as Edison and Bell, there was a change. With economies of scale such as Ford achieved with serial mass production, and as giant corporations rapidly came to dominate markets, professional management was needed [LAC 87]. When Will Durant took over what remained of the US auto mobile industry in General Motors, banks and shareholders insisted on bringing in Sloan, an accountant rather than an engineer to manage, which developed into 3M (Minnesota, Mining, and Manufacturing Company Limited) multidivisional management structure, which then became paradigmatic for big business [SLO 64, WOM 05, WOM 96].

But in the process a divorce emerged. Veblen [VEB 21] in his *The Engineers and the Price System* wrote that "businessmen are increasingly out of touch with that manner of thinking and those elements of knowledge that go to make up the logic and the relevant facts of mechanical technology…". He allowed that industrial experts, engineers, chemists, mineralogists and technicians of all kinds have been drifting into more responsible positions in the industrial system and have been growing up and multiplying within the system, because it will no longer work at all without them [VEB 21, p. 26], but they rarely became professional managers. Veblen also castigated emerging neoclassical theories of the firm for their presumption that managers simply combined capital and labor and that anything else, such as technical progress was simply a residual. In contrast, like Schumpeter [SCH 49], he claimed that it was engineers who created new products, processes and, with them, enabled new markets, while Schumpeter himself contrasted creative innovators from managers, as administrators. Corporate strategy became important [CHA 62], yet led to inflexible mind sets [SEN 90] and then declined in effectiveness [MIN 94] such as that there was nothing to be learned from post-Fordist models of flexible production and "lean" engineering developed in Japan [WOM 05, WOM 96] which is now common in Asia.

This relates to innovation trajectories and product cycles. For decades, in western management theory, it was assumed that a product is first innovated, then grows and then matures [VER 66]. Yet, according to Schumpeter [SCH 49] this may spawn other products that may transform the initial innovation, such as mobile phones initially displacing fixed line telephones, then incorporating cameras and keyboards and enabling access to texts without the need to go to a library. Such transformations were achieved by engineers and it is engineers who can advance this unless they are blocked by managers who may have no training or experience in engineering [OLI 13].

Perceptions of roles and identities also are influenced by either explicit or implicit logics [OLI 07a, OLI 06, OLI 00, OLI 12]. A profession such as engineering has distinctive features similar to medieval guilds [DUR 57] including a formal code of ethics, criteria for certification as member of a recognized professional association and the monopoly of a specific labor market by regulating the entry of members. By contrast, management does not have these features. Even if there are explicit commitments to ethical and other codes in corporate mission statements, these may prove to be rhetorical [LOK 10]. Tacit rules and implicit norms in organizations [OLI 07a], whether as in Enron, World.com or the more recently revealed Libor interest rate manipulation, may override the explicit commitment to ethical codes in mission statements. Engineering and management also have different criteria and sanctions for performance. If an engineer designs a bridge that collapses, then his or her professional reputation is destroyed. If bankers design financial derivatives that collapse the western world, they can continue to pay themselves bonuses unrelated to performance, which may change public perceptions of banks, and incur public protest, but without necessarily affecting their careers.

Barley [BAR 06] has cited a range of studies distinguishing between basic and applied researchers. These indicate that the latter are usually engineers who are more concerned with recognition within organisations and a managerial career, unlike scientists and management theorists for whom recognition depends on publishing research findings. Yet these, and perceptions of them, may be asymmetric. An engineer working as manager is likely to be accepted with or without a professional qualification in management whereas this is not the case with a manager presuming to be an engineer.

This chapter indicates that engineers nonetheless may be reclaiming the roles and professional identities of earlier engineer-managers. It analyzes this in terms of theories of personal and professional identity, subject–object relations in terms of theories of "the self and the other" [SCH 18, SAR 57, HOL 13], the role of self-direction and other direction [RIE 54] and factors influencing both initial career choice and later career moves in relation to concepts of boundaryless or Protean careers. In doing so, the chapter extends Schneider's [SCH 83, SCH 87, SCH 90] attraction–selection–attrition (ASA) theory or ASA model from personnel selection and candidate choice to career choice and draws on but qualifies Foucault's [FOU, 72, FOU 77] concept of power-knowledge. Using case study evidence, it illustrates the mediating effect of the perceived comparative status (PCS) of engineers toward managers in the relationship between professional identification and professional commitment.

In asking whether managers are a professional reference group for engineers and, if so, whether that influences career choices, the chapter addresses the following exploratory propositions (EP):

EP1: engineers perceive cultural differences along with an unbalanced status between engineers and managers;

EP2: Foucault's theory of power-knowledge can be qualified in terms of managers having power but engineers having knowledge;

EP3: Schneider's ASA theory is relevant to engineers initially being attracted to engineering as a profession but then wishing to avoid narrow job-fit by qualifying as managers;

EP4: it has implications for differing perspectives on boundaryless or Protean careers.

1.2. Identities and values: the self and the other

The relationship between engineers and management has been studied in terms of different roles and identities, as well as different organizational cultures ranging in Human Resources Management (HRM) terms from "soft" in terms of cooperation and group working in Japan and other Asian cultures to "hard" in some more individualistic market cultures such as those of the United States [BOW 10, LEE 92, LEE 06, LEE 10, SHA 98, SHA 03]. This varies in whether or not there is a high degree of trades unionism, including professional associations [STU 08], even if some professions prefer not to conceptualize their associations in such terms, as well as with different degrees of commitment of higher level management to human resource management policies such as retraining and requalification [CAB 04].

Identities also involve different perceptions and cognitions that are influenced not only in terms of a Cartesian *cogito ergo sum* [DES 37] but also in terms of who we have become through the acquisition at varying levels of consciousness of values, beliefs and dispositions derived from both life and professional experience [HOL 13]. Such cognitive and social processes are not only functional but also existential in terms of relations between "the self and the other" in the sense of existentialist philosophy and literature from Schopenhauer to Sartre [SCH 18, SAR 57, HOL 13], where the self and individual perceptions are subjective and include choice and scope for self-direction whereas the other is objective, as in an organization or a profession.

This self–other distinction is paralleled in the cognitive-experiential self-theory of Epstein [EPS 90, EPS 91, EPS 92, EPS 93, EPS 94] that premises that "...everyone develops an implicit theory of reality that contains subdivisions of a self-theory, a world theory, and propositions connecting the two" [EPS 90, p. 165]. This relates also to whether people are disposed to be self-directed or other-directed that Riesman [RIE 50, RIE 54] developed in relation to whether they are disposed to be "one of a crowd" or to stand out, or strike out on their own.

Cognitive psychologists have wavered between the view of the individual as a competent and rational being and a view with limitations for rationality and its biases [SCO 01]. Identity is influenced by mental pattern and schemata in a categorization system with group influences, expectations and even stereotypes [TUR 87, BAR 95, HOW 00, SEN 90]. Burke [BUR 04] explains the relation between the individual and the social structures with the identity control theory. Identity theories also recognize differences between passivity and non-passivity of the individual in relation to contextual factors, social systems and roles [BUR 04], which parallels Riesman's [RIE 50, RIE 54] concepts of self-direction or other-direction. Careers relate to institutional logics, mechanisms of corporate governance and division of labor [JON 07]. However, this is not always in a manner that may be self-enhancing or career enhancing, as in how Adam Smith himself recognized that division of labor may "numb the mind" [SMI 76, ROT 00] and the degree to which this may incentivize engineers to escape this by seeking to qualify as managers rather than be constrained as "mere" technicians.

Careers therefore involve both subjective and objective factors. The subjective concerns perceptions, senses and feelings by individuals about their own career and their future while the objective includes perceptions by others [ARN 08, NIC 05]. This ontological duality is recognized, notably by the Chicago School, where the objective dimension is associated with forms of professional identity in society and the subjective dimension is represented in the meanings that individuals attach to their career, while these may change over time [BAR 89]. A career can be planned in advance, such as choosing in the first instance to study engineering and becoming an engineer, yet then retrospectively reconsidered [YOU 92, YOU 04] also in terms of seeking greater personal fulfillment through a career change.

Both subjective and objective factors profoundly influence careers and career choice [ARN 08]. Some of the variables are also more tangible than others [VAN 77a] while some are social in terms of shared perceptions of what is a good career or a good career choice [ART 05]. Nicholson and de Waal-Andrews [NIC 05] have suggested six recurrent career indicators including: (1) hierarchy and status, (2) material success, (3) social reputation, prestige and influence, (4) knowledge and skills, (5) personal relationships, contacts and friendships and (6) health and well-being.

Identity is formed and sustained by social interaction. It allows us to see ourselves as similar to a class of individuals with whom we associate ourselves or want to be associated [GIO 98, TAJ 86, MAT 88]. Self-categorization is integral to social identity [ASH 89]. People tend to assume identities in relation to others in prototypical terms such as belonging to a particular organization, gender identities and religion [ASH 89]. It plays two roles: both cognitive at varying levels of consciousness and as an ordering of the social environment. Howard [HOW 00] stresses both individual cognition and social interaction in the development of the self. Burke [BUR 04] relates this to validation and verification, assuming that once an identity is adopted it must be continuously controlled for its social validation, while internalization of acceptable behavior and symbols is part of its self-verification.

Identity, self-direction and career development may be resource constrained, ranging from whether an individual has the resources to invest in a further qualification, such as an MBA or MSc, to whether a general economic environment is constrained by lack of market demand and there are pressures on governments to reduce public spending and investments, as since the onset of the Eurozone crisis in which one of the first investments to be cut or postponed may be major engineering projects [VAR 12]. Yet, Tajfel and Turner have put forward a theory of social identity independent of resource constraints [TAJ 79, TAJ 86]. This recognizes the importance of groups in individual identity such as the difference between in-group and out-group identities that are not necessarily economic or material. The reasons given for this are (1) the need for positive self-esteem, (2) self-perceptions within particular groups (social identity), (3) a positive social identity that can be achieved or enhanced by comparison with relevant out-groups and (4) differentiation in terms of identification *foci*.

The salience of social identity is a dynamic variable depending very much on organizational and operational context [DIC 05]. For example, the knowledge and skills important for professional practice are necessary but not sufficient conditions for professional preferment. As with Riesman's

[RIE 50, RIE 54] concept of self-direction, it favors those who have both a strong personal and professional identity [COS 05, p. 29], yet also implies that they should not be so self-directed as not to disregard the tacit rules and implicit norms of a group [OLI 07a], especially at operating levels such as in a legal or design practice or an engineering group. Professional labeling has been claimed to be a mechanism for control in the sense that it constructs the job identities and appropriate behaviors [FOU 99]. The meaning of being professional is therefore an important discursive construction [CLA 09], but not only so. Professional identity is sensed and felt rather than necessarily being explicit [MAT 88].

1.3. Symmetries, asymmetries and career dynamics

Perceptions by individuals, groups or professions may be either symmetrical or asymmetric. For example, an engineer working as a manager is commonly accepted. The presumption that a manager can take engineering decisions is not accepted. Symmetries and asymmetries also relate to both individual and group identities and to both sets and subsets of meanings at varying levels of consciousness [MAT 88]. This is an ongoing process of symmetrizing who we are and what we are expected to be, or what we perceive that others want us to be [MAT 88]. It is integral to the principle of semantic congruence [BUR 81] and thus gaining consistency [HAV 94].

Such processes are dynamic [DUB 96]. The interaction concerns space for the self and for social interactions with others [MAC 03], yet may not be linear rather than dialectical, such as finding that there is asymmetry rather than symmetrical "fit" between what we aspire to be and what is expected of us by others and then seeking a career move. Or that a young engineering graduate may initially be fulfilled in gaining a "good job" and glad to be able to sustain it, whether or not a partner is working, yet in midlife sense that they are "ceilinged in" by a narrow job-fit, which may mean that they seek to qualify as managers and achieve higher levels of self-realization in terms of Maslow's hierarchy of needs [MAS 43].

Engineers who start their careers as such may come to want more than simply working as one, and seek a career shift into management, even if the prospects for this remain uncertain [BAK 96, ROU 98]. Whether they do so by Hirschman's "voice" internally with an organization, or by needing to "exit" [HIR 70] will be influenced by not only by their own sense of self, but also by how they are perceived by others at higher levels of management [BAK 96, MIR 94, KIN 05].

John Arnold [ARN 01] stresses that careers imply subjective and objective interactions, potentials and constraints. A career is a sequence of positions, roles, activities and experiences encountered by an individual, whether employed or self-employed. Career progression is not limited to upward mobility within an organization. Horizontal mobility may be as important if it means job variation, while stepping back or down, as in opting to work part time, is also a career move, which is also supported by others [ART 96, ART 05, OLI 07b]. Lee [LEE 92] analyzes dynamics in the relations between management and engineering that vary between different cultures, often with overlapping roles.

Cultural context should not be displaced, yet nor should it be exaggerated. For example, the engineering breakthroughs that were achieved by Toyota after WW1 in achieving *kaizen* style continuous improvement were not dependent on a Confucian culture but on a very western style industrial dispute [WOM 96] as well as by the *kanban* principle of automatic restocking and delivery of components on a just-in-time rather than just-in-case principle that had been developed earlier by American retailers yet never taken up by US manufacturers [COL 00a].

In addition, line with shifting dynamics in individual or group perceptions, some objective indicators of success such as pay and position may gradually lose their significance in terms of Maslow's hierarchy of needs, and higher levels of self-fulfillment become more important [MAS 43]. Thus, what are considered career needs may change over time [ART 99, ART 05]. When a social identity is unfulfilling, psychologically or economically, either in terms of job-fit, group-fit or organization-fit, the relationship is asymmetric and an individual may exit with the aim of joining or forming others with more symmetrical values and goals. Yet career change need not imply "exit" rather than "voice" [HIR 70]. For managers whose skills are in market demand, "voice" may enable recognition, with or without formal negotiation, and promotion with associated pay and other benefits [ART 05].

Professionals can find stability in an ambiguous environment using anchors of identity such as being expert or being different [BEE 08]. Representations (physical, symbolic, verbal, textual and behavioral) become imbued with meaning and are taken as part of the identity [BEE 08]. Identity has multiple dimensions of which three are: (1) the values, dispositions and beliefs that individuals implicitly acquire not only from professional but also life experience [OLI 07a, OLI 13]; (2) tacit rules and implicit norms in behavior, which may both underlie and override formal job descriptions or organizational mission statements [OLI 07a, OLI 13] and (3) discourse in identity production and reproduction [BEE 08].

In acquiring and defining social and professional identity, individuals may consciously seek professional qualifications and also accept the norms that are embodied in work roles [DUT 94, ELS 01, HUM 02]. This may symmetrize self-esteem with the identity of the organization including its external image and depend on whether an individual is identified as "fitting" within a presumed identity profile [ASH 90, ASH 89, DUT 94, HIN 89, HUM 02] or as "one of us" in the famed term of Margaret Thatcher, which could imply that they either are included and may gain one or more rungs on a promotion ladder, or excluded as permanent "outsiders".

Professionalization is therefore a dynamic process through dyadic pass or fail for entry to the profession, followed by explicitly defining criteria for work methods [DIM 83] through tacit rules and implicit norms acquired at varying levels of consciousness from work experience [OLI 07a]. This requires symmetrization [MAT 88] including "fitting in" rather than only the knowledge, abilities and skills for job-fit [OLI 07a]. Yet the need to symmetrize with the image of an organization may also provoke dissent rather than assent, with ridicule of prescribed roles in camaraderie and humor, counter-narratives, irony and cynicism [CLA 09]. Within some organizations individuals and groups may have considerable freedom in creating their reality, shaped by the social discourses available [HUM 02]. Identity is an interactive process equivalent to the writing of one's history, seeing how it be written by others and identifying ourselves in the discourses of others [SIM 05a, SIM 05b]. On the other hand, it may be a process of absorption within a larger narrative [BEE 08].

1.4. Evolving identities and professional reference groups

Identity formation by reference to others is central to self-development from childhood in terms of parental or other role models, peer groups in adolescence and to professional groups. Some reference groups are stable, durable and long lasting [KRO 76]. Others may change either slowly, or quickly, such as in the case of perception of breach of psychological contract [ROU 95, ROU 98]. Relationships also tend to imply power [KRO 76] and power-knowledge may be implicit at any level of an organization [FOU 72, FOU 77]. A professional group that holds hierarchical power does not need to change its values or terms of reference [KRO 76]. Also, although self-identity is largely formed by late adolescence or early adulthood, it may be changed by less than conscious referential processes in operational contexts and by language specific to a profession or work group [INK 07].

Thus, group and professional identities are not static but interact dynamically [TAJ 79]. The knowledge and narratives shared by group members influence choices and actions and less consciously defined boundaries. Lipiansky distinguishes between belonging to a group, linkages between groups and reference groups influencing behavioral values and norms [LIP 90]. Identities are therefore internal to an individual as "self" yet also relate the individual to "the other" [SCH 18, SAR 57] and interface with different social situations, contexts and relationships, and change over time [BUR 98].

Professional identity construction may start in families, as with children of doctors becoming doctors, but will also be influenced by secondary and higher education choices [DUB 97]. Later socialization and training in work contexts have a key role [MIG 09]. Professional identity is based on a sense of belonging to a professional group recognizable by society [PER 07]. Socialization at work is the product of power relations and collective standards for behavior, with a representation of identity as a backdrop. Professional identity is linked to the adoption of norms and values of a particular profession. The strength of an identity can be related to the commitment to the profession and to the understanding that the organization represents a space for the advocated professional identity. As Luckmann and Berger [BER 73] emphasize, the development of a professional identity is assumed as a dimension of secondary socialization, applied in the specific field of an activity with specific symbols. In addition, identity is constructed from expected roles [BAU 98].

As supported by the findings of Burke and Franzoi [BUR 98], identities are dynamic carriers of meaning and behavioral expectations. Professional identity is confirmed in the process of self-realization and verification [DOW 09] or by falsification of initial expectations, or changed expectations, similar to Schneider's [SCH 83, SCH 87, SCH 90] ASA theory. Thus, identity is formed by the dynamics of social interaction [GIO 98]. Identity works as a distinctive self-classification and simultaneously allows us to see ourselves as similar to a class of individuals with whom we associate or want to be associated [GIO 98, TAJ 86]. In line with this, Howard [HOW 00] stresses both social cognition and social interaction. Once an identity is adopted, it may be dynamically reproduced or reinterpreted [HOW 00].

People may simultaneously identify with multiple identities [THA 08]. They are not necessarily identified with all the groups to which they belong or to all the roles they are called on to play. Identification can have multiple *foci* [DIC 04, HAR 05, KNI 00]. These can be centered on the career (at the individual level), the team, the organization or occupation (social level) [DIC 04]. Different processes in identification can include: (1) cognitive (knowledge of membership of a group), (2) affective (emotional attachment to a group), (3) evaluation (evaluative connotations both internal and external) and (4) different or similar behaviors [DIC 04]. Identification and commitment have different geneses: identification is contingent on the basis of a shared destination and perceived similarities with the organization, whereas commitment may be transactional, that is the material relationship between the individual and the organization [DIC 04]. When there is a change in context or in a comparison, individual or group identification can recalibrate [BAR 01, HOG 00].

A professional image is a mirror of professional identity, despite its complexity. Identities emerge from the interactions, negotiations and production of meanings, mediated by organizational contexts and interactions with internal and external audiences [CLE 07]. This process is one of creation rather than an ontological security. It defies stasis and predictability [CLE 05, CLE 07]. Career experiences provide the individual with objects and people that can lead to the stabilization of self-perceptions of identity or transformation into new identities [BEC 56]. Among the mechanisms that affect the participation experience and, consequently, the identity, we find: (1) the development of interest in the problems of the profession and pride in new skills, (2) the acquisition of a professional ideology, (3) internalization of motives and (4) the investment (level of effort in acquiring skills). Identification with work is affected by the following conditions: (1) connection to occupational title, (2) commitment to the tasks and (3) commitment to organizations or positions held [BEC 56].

1.5. Protean or boundaryless careers

Research on careers provides a bridge between cognitive and social levels of analysis [JON 07]. Institutional processes encode social knowledge of roles and role relationships. Then, these may come to be collectively shared and survive across generations [ZUC 77, JON 07]. Yet, these depend on perceivers, and may either be false or, in line with Popper's falsification principle [POP 59, OLI 12], may prove to be false over time, even if there is a lag between this occurring and either individuals or professions themselves recognizing it. This also relates to perceptions and misperceptions, depending on less than conscious values, dispositions and beliefs [OLI 12, HOL 13].

Thus, there has recently been a disposition to presume that increased competitive pressures with globalization make workers more responsible for their path and necessarily proactive and adaptive [RAA 07]. This relates to two seminal metaphors in career literature: (1) the boundaryless career and (2) the Protean career. Collin and Young [COL 00a] submit that the concept of a boundaryless career is about flexibility, autonomy and self-determination. Brocklehurst focuses the notion of a career without boundaries in the discussion of identity, but questions the extent to which this is important in establishing or extending the sense of self [BRO 03].

Inkson [INK 04] has stressed limits to career metaphors while, in commenting on the metaphor of a "Protean" career, Arnold *et al.* [ARN 08] have observed that Proteus changed shape not by choice but to avoid capture, which also may be an increasing sense of being captured in either a boring, routine and alienating job or one that an engineer realizes is not fulfilling his or her potential since constrained by narrow job-fit rather than wider and higher level organization-fit. Thus, on the one hand, concepts of boundaryless or Protean careers may imply that a self-directed individual assumes control or that they have been constrained to change jobs either by the termination of a contract or by being fired [HAL 76]. Our approach relates this to the dynamics of Schneider's ASA theory in career choices [SCH 83, SCH 87, SCH 90]. Thus, we suggest that while someone may initially be attracted to select engineering as a career, its attraction may wane with time and that of management increase.

Heslin [HES 05a, HES 05b] has stressed change over time in terms of perceptions of career success. Thus, success can be at a particular moment, such as winning a contract, or over a longer time period, such as consistently doing so. Nicholson and de Waal-Andrews can also be defined in terms of personal recognition, or interpersonal comparisons, or other forms of success such as income [NIC 05]. Subjective evaluation of career success relates both to perception of others and a process of internalization [ART 05, VAN 77a]. Different social actors have different career aspirations and different valuations of success in terms of security, place of work, status, promotion, access to further education and training and work-life balance [NIC 05].

These also differ in relation to objective factors such as gender and ethnicity, whether the individual is a manager, a secretary or a technician and it therefore would be mistaken to assume that all members of a given social or employment category share the same subjective career perceptions [BAI 89]. Moreover, there may be gender differences such as women preferring subjective evaluation of success whereas men prefer it in the perceptions of others such as recognized status and preferment [HES 05a]. This may also relate to the labor market. Highly specialized individuals with scarce skills may be more disposed to opt for career mobility when they sense or know that what they can offer is in demand [ARN 08]. This can also concern identity "regulation". Alvesson [ALV 01] has claimed that this relates to how employees are expected to govern themselves and to accept that they should be governed, such as: (1) identification with the organization as an institution, (2) cultural control and (3) standardized norms of behavior.

Governance implies power relationships and also merits a distinction between strategy and tactics. Some of the literature on boundaryless careers has conceptualized this as a positive strategic choice by individuals, whereas it may be a tactical reaction to negative experience. We suggest that this, and Schneider's ASA theory model [SCH 83, SCH 87, SCH 90], can be informed by Foucault's [FOU 72, FOU 77] case that power relationships can lie at any level within institutions and prove oppressive; such as a lower level manager with little to no knowledge of engineering assuming the right to command and instruct engineers in what may be a technical decision. With the qualification nonetheless of Foucault's case of power-knowledge in the sense that managers may have the power to support or inhibit innovative engineering, yet without engineering knowledge, whereas the engineers have the technical knowledge but lack the power to apply it and thus to innovate either products or processes according to Schumpeter [SCH 49], unless they become managers. Thus, they may wish to transcend constrained perceptions of job-fit by gaining professional qualifications as managers in the hope of being able to apply their own experience at higher operational and organizational levels.

Despite increasing recognition of the need for work-life balance [HIL 06], this may also constrain allegedly boundaryless or Protean career choices. Being either married or in a relationship with small children tends to imply more concern with family time, while this is also constrained by location such as needing not to work too far from schools [KIR 06] that also may mean forgoing a promotion. This is less the case for men, even if with downsizing and outsourcing they then may face the risk of being unemployed in mid-career, while women may then be at a double disadvantage in the sense that while they may still have a job, they may be on a "career plateau"

[KIR 06]. Whereas, on the other hand, there is some evidence that not having children, or being single, is associated with less career success for both men and women [THA 99] while, although pay is not the only factor in career choice [RYN 04], it cannot be disregarded.

1.6. Dialectics, dilemmas and career choices

In line with a dialectic between subjective and objective factors, Schein has qualified the concept of an “organizational career” and proposed instead the concept of an “internal career” as a sequence of experiences that may be linear, with an internal sense of being “anchored” and secure, or may be fragmented [SCH 96], while it may also be that “career anchors” and a sense of being “embedded” may be satisfied with what one has rather than risking a major career change, such as re-qualifying for a different profession. This suggests that the concept of a career, and of career choice, also needs to take account of the interdependence of subjective perceptions and objective indicators of success, such as not only wanting a wider range of experience and responsibility, and to “move on” but also realizing that one has reached a ceiling within an organization and has nil or near to nil chance within it of being able to “move up”.

Nonetheless, some researchers continue to assume that career success relates to upward mobility on an internal career ladder [ART 05]. This tends to assume not only stable careers but also stable organizations, whereas their stability with globalization, downsizing, outsourcing and business re-engineering for some time has been in question. Moreover, according to Arnold and Cohen career success depends both on performance and the social position of the individual [ARN 08]. Claiming support from Judge *et al.* they submit that there is more evidence supporting the role of social position than that of performance [JUD 04]. Pay levels and promotions relate to objective success, whereas subjective success concerns career satisfaction [JUD 04].

Doyle [DOY 00] has stressed the need to avoid mechanistic models of career management in favor of more holistic and integrated approaches. According to London [LON 93], career motivation is a multidimensional construct embodying at least three factors: (1) capacity to adapt to different circumstances and acceptance of risks; (2) realism in terms of clear objectives related to self-knowledge of one’s own capacities and challenges; (3) identity, in the sense of the degree to which an individual can be symmetrical with both the demands and needs of a job and of the organization.

There is also the distinction between know-how and know-whom. Know-whom is important both for traditional and boundaryless careers, and in smaller institutions such as family firms, whereas know-how is more job related [ART 95]. Table 1.1 represents the comparison between traditional and Protean careers [DEF 94, HAL 04].

	Traditional career	Protean career
Boundaries	One or two organizations	Multiple organizations
Identity	Employer dependant	Employer independent
Relationship	Job security	Employability for performance and flexibility
Competences	Organization specific	Transferable
Measures for success	Salary, promotion and status	Psychologically meaningful work
Career management responsibility	Organization	Individual
Key attitudes	Organizational commitment/involvement	Work satisfaction and professional commitment/involvement

Table 1.1. *Adapted from [DEF 94] and [HAL 04]*

Self-directed careers can be of three types: (1) positioning, through accumulated contacts and experiences, intended to yield a given result; (2) gaining influence with other members of an organization and (3) opening new frontiers [KIN 04]. According to Kanter's concept of a portable career, a successful individual can carry or transfer this either elsewhere in an organization or to another organization [KAN 89]. Arnold and others have also seen careers in terms of a theory of action planning [ARN 06, RAA 07], related positively with career satisfaction [RAA 07].

Ashforth and Mael [ASH 89] identify four factors related to group identification that may be related to career embeddness or career anchors: (1) group values and practices, (2) group prestige and the degree to which this may enhance for an individual, (3) in-groups and out-groups in terms of social frontiers and (4) group dynamics in terms of interpersonal interactions, similar values, the sharing of aims or responses to threats.

1.7. Case study, sample and data analysis

A sample of 315 engineers voluntarily answered a web-based online questionnaire survey. They all belonged to the Portuguese Society of Chartered Engineers, affiliated with the Portuguese Centre Regional Committee. There were no previous restrictions or cohorts. The mean age was 36.74 years (standard deviation (SD) = 8.99 years). The sample consisted of 51.4% female respondents and 48.6% male respondents. All the respondents worked at the time the questionnaires were answered. However, 13% did not work as engineers. Of the participants, 9.3% were self-employed, 84.5% worked on behalf of others and 6.3% worked as independent workers or had mixed situations. Of the respondents, 47.6% said they held a management position and 52.4% did not have a management position.

The questionnaire items evaluating the research questions were developed from the existing research. The items were translated into Portuguese by the research team and back-translated to English by a professional translator to check for bias. The pre-testing of the questionnaire was made online and it was applied to the targeted population; data analysis was done in the Statistical Package for Social Sciences (SPSS) 21 software. We adapted the professional identification (PI) scale developed by Hekman [HEK 09] by substituting doctors for engineers and reversed the scale (1 = "disagree completely" to 5 = "agree completely"). The professional commitment (PC) scale developed by Meyer *et al.* [MEY 93] consists of an 18-item measure of affective, continuing and normative occupational commitment (six items in each scale). The items were modified to apply to the engineering profession (1 = "strongly disagree" to 7 = "strongly agree").

The engineers and managers relationship scale (EMRS) was previously used in studies comparing engineers and marketers [SHA 98, SHA 03]. Eleven items were used and modified to apply to the engineering profession compared to managers in general. Participants responded to each of the measures using a five-point-type Likert scale ranging from –2 (strongly disagree) to 2 (strongly agree). The perceived comparative status (PCS) scale comparing managers and engineers was created using eight variables: (1) salary, (2) social status, (3) amount of work, (4) career perspectives, (5) job security, (6) job interest, (7) job satisfaction and (8) autonomy (with a five-point scale ranging from much better, better, equal through to worse and much worse).

1.8. Results

1.8.1. *Professional identification (comparison group)*

When questioned about the professional group suitable for comparison with engineers' work and careers, 32.71% of the respondents referred to managers. References to other engineers come second to references to managers (Table 1.2).

Professional group for comparison*	Valid (%)
Managers	34.65
Other engineers from other sectors	14.52
Architects	14.52
Lawyers	11.88
Doctors	8.58
Public sector employees	6.60
Economists	4.62
Others	4.62

*Valid percent excluding "none" = 12.53% and "all" = 4.90%.

Table 1.2. *Professional groups mentioned by comparison*

1.8.2. *Mediating effect of the perceived comparative status*

The concept of mediation implies the assumption of causal relationships between the variables involved. A mediator variable is one that in the regression equation decreases the magnitude of the relationship between an antecedent variable and a dependent variable or criterion. There are both scope and limits to this. Dunlap and Landis [DUN 98] have found that building internally consistent measures that include multiple dimensions correlated with the same construct can lead in some cases to excluding important variables in explaining the focused variable. In these cases, it is safer to use statistical techniques such as multiple regression.

We performed a hierarchical multiple regression to describe the relation between PI and PC and study the mediating effect of the PCS. The beta scores (Figure 1.1) indicate that PCS partially mediated the relationship between PI and PC.

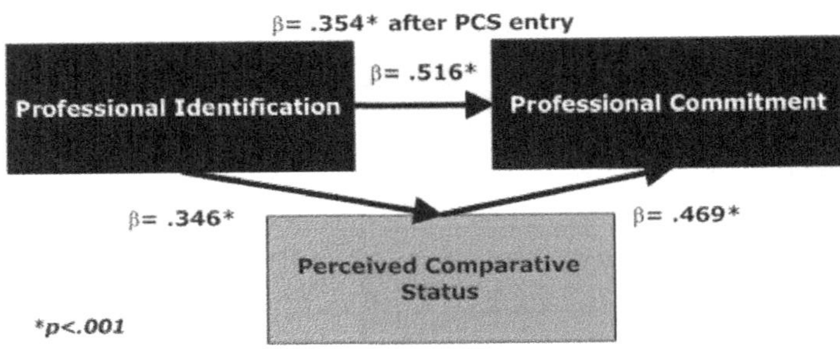

Figure 1.1. *PCS mediating effect on PI and PC*

Recoding the PCS construct in negative, equal and positive and using the one-way analysis of variance technique for comparing the PC provided the following results: the Levene's test of homogeneity of variances (2.601; sig = 0.076) enabled the use of the Scheffe's *post hoc* analysis (Table 1.3):

Dependent variable: professional commitment

Perceived comparative status	Mean difference	Significance	95% Confidence interval		
Negative	Equal	–0.801	0.000	–1.193	–0.408
	Positive	–0.534	0.000	–0.727	–0.342
Equal	Positive	0.266	0.292	–0.150	0.683

Table 1.3. *One-way ANOVA of professional commitment segmented by the recoded PCS*

The mean difference between the PC of those engineers who feel a negative comparative status is statistically significant from those who feel the status as equal or positive (favoring engineers). This adds confidence on considering PCS a mediator of PC.

The recoded PCS construct reports 72.7% of the respondents as feeling negatively affected (Table 1.4).

		Frequency	Percent
Perceived comparative status	Negative	229	72.7
	Equal	14	4.4
	Positive	72	22.69
	Total	315	100

Table 1.4. *Isolated variables of the PCS*

The analysis of the isolated variables of the PCS construct is also interesting, as shown in Table 1.5.

	Negatively unbalanced (%)	Neutral (%)	Positively unbalanced (%)
Salary	**83.5**	13.3	3.2
Social status	**50.5**	39.1	10.4
Workload	37.5	**48.9**	13.6
Job security	37.1	**38.1**	24.8
Interesting job	13.7	35.9	**50.4**
Work satisfaction	19	**54.7**	26.3
Autonomy	29.5	**48.6**	21.9

Table 1.5. *PCS (frequency table)*

Salary and social status are the most negatively unbalanced indicators. Having an interesting job is considered positively balanced by a short majority of the respondents (50.4%).

1.8.3. *Management training/education*

Testing for the independence of the variables (1) having had training in management and (2) being a manager results in a Pearson Chi-square of 0.256 (sig>0.05), suggesting that not only actual managers have had the training. Of the overall respondents, 35.2% have had some form of training or a course on management and 86% of them did this through personal initiative.

Management training/education	Decision maker	Reason
Yes (35.2%)	Me (94.6%)	Develop competences for the current job (23.16%)
		Develop competences for a new job (76.84%)
	My employer (5.4%)	Develop competences for the current job (93.75%)
		Develop competences for a new job (6.25%)
No (64.8%)	–	–

Table 1.6. *Management training/education, decision and reasons*

The results (Table 1.6) indicate that 35.2% of the engineers have had training or a management course and that this was done mainly by personal initiative. The reasons for entering a course in management are completely opposed when segmented by the decision-maker: when the decision is made by the employee, it is intended for a new job and thus related to an effective career move; when the decision is made by the employer, it is intended for the developing of competences in a current job.

1.8.4. *Relationship between engineers and managers*

A principal components analysis with a varimax rotation was performed on the engineering and management relationship scale (EMRS) in order to explore the data (exploratory factor analysis (EFA)) and assess the construct validity of the study measures. The subjects to variables (STV) ratio is >20:1 (n = 315 and variables = 11) permitting the use of this statistical option.

Results of the principal components analysis yielded four factors (factor 1: eigenvalue = 3,184; factor 2: eigenvalue = 1,984; factor 3: eigenvalue = 1,307 and factor 4: eigenvalue = 1,168), explaining 69.51% of the total variance. The measures for sampling adequacy (MSAs) indicate the possibility of using the data for EFA (KMO = 0.664; Bartlett's test of sphericity sig = 0.000).

The interpretation of the factors is shown in Table 1.7.

Factor	Variable	Loading	Interpretation
1	Cultural differences exist between engineers and managers	0.852	Communication and culture
	It is difficult for engineers to communicate effectively with managers	0.656	
	Cronbach's alpha = 0.654		
2	Engineers and managers should work together to solve problems	0.804	Engineering and management integration
	Teamwork between engineers and managers should be encouraged	0.740	
	Engineering and management should be closely integrated	0.605	
	Management and engineering are two completely separate functions	0.558	
	Cronbach's alpha = 0.696		
3	Engineers and managers do not understand each other	0.825	Status and conflict
	Engineers are more important to a company than managers	0.736	

Table 1.7. *Factors and interpretation*

	Engineers and managers should have equal status	0.637	
	Cronbach's alpha = 0.617		
4	All engineers should know something about management	0.750	Knowledge and trust
	Engineers and managers cannot trust each other	0.652	
	Cronbach's alpha = 0.226		

Table 1.7. *(Continued) Factors and interpretation*

There are four interpretable factors: (1) communication and culture, considering the cultural differences and communication issues; (2) engineering and management integration, considering the acknowledgement of two different disciplines that can be integrated; (3) status and conflict, including the power relations and role importance; and (4) the relationship between knowledge and trust.

When analyzing the descriptive statistics of variables like the perceived cultural difference (Table 1.8), we find that 65.4% of the respondents acknowledge such differences, 21% are neutral and only 13.7% disagree. Interestingly, when segmented by those who are currently in a managing position or not, engineer-managers perceive this difference even further, with a relative frequency of 81.4%.

The variable "it is difficult for engineers to communicate effectively with managers" also has interesting results in terms of differences between engineer-managers and engineers who are not managers. Of the engineer-managers, 28.7% consider that effective communication is not difficult; on the contrary, 44.8% of the engineers who are not managers consider the difficulties to be true.

Table 1.9 shows that more than four-fifths of the respondents (84.4%) disagreed that management and engineering are two completely separate functions. Nonetheless, 91.7% of the respondents also disagreed with the statement "All engineers should know something about management". This high number diverges from more than one-third of engineers seeking training in management in order to qualify for a new career.

Cultural differences exist between engineers and managers		Frequency	Percent
Currently in a direction or management position?	Totally disagree	1	0.7
	Disagree	12	8.0
	Neither agree nor disagree	15	10.0
	Agree	97	64.7
	Totally agree	25	16.7
	Total	150	100.0
Currently not in a direction or management position?	Totally disagree	3	1.8
	Disagree	27	16.4
	Neither agree nor disagree	51	30.9
	Agree	68	41.2
	Totally agree	16	9.7
	Total	165	100.0

Table 1.8. *Frequency table for the variable "cultural differences exist between engineers and managers"*

		Frequency	Percent
Management and engineering are two completely separate functions	Totally disagree	87	27.6
	Disagree	179	56.8
	Neither agree or disagree	16	5.1
	Agree	29	9.2
	Totally agree	4	1.3
	Total	315	100.0

Table 1.9. *Frequency table for the variable "management and engineering are two completely separate functions"*

1.9. Discussion and conclusions

Exploratory proposition 1: engineers perceive cultural differences along with unbalanced status between engineers and managers.

The data in the case study support the proposition. Cultural differences in status are acknowledged by 65.4% of the overall respondents and 81.4% of the engineer-managers. These data suggest a deeper assumption that professional identities are different and those engineers who have a management position reinforce the idea that they have entered a different world where probably some professional identity redefinition is needed. The PCS construct highlights the concept that engineers feel underprivileged in accessing objective (or indirectly subjective) indicators like: (1) salary, (2) social status, (3) amount of work, (4) career perspectives, (5) job security, (6) job interest, (7) job satisfaction and (8) autonomy. Salary and social status are the most negatively unbalanced indicators (83.5% and 50.5% of the references, respectively).

Exploratory proposition 2: Foucault's theory of power-knowledge can be qualified in terms of managers having power but engineers having knowledge.

This relates to the proposition that an engineer can become a manager but a manager cannot become an engineer. Results show that 35.2% of the engineers in our sample perceive the need to formalize management knowledge by entering management courses, and this intended behavior is self-directed in the sense of Riesman [RIE 50, RIE 54] and aimed at a future career change. The analysis shows that there are four related factors: (1) perceptions of status, (2) engineering and management integration, (3) role importance and (4) divorce between knowledge and power. This qualifies Foucault's [FOU 72, FOU 77] equation of knowledge and power in the sense that it is engineers who have technical knowledge yet managers have power, including the power to support or inhibit innovative engineering [OLI 13].

Exploratory proposition 3: Schneider's ASA theory is relevant to engineers initially being attracted to engineering as a profession but then wishing to avoid narrow job-fit by qualifying as managers.

More than one-third of the respondent engineers perceived the need for training in management. It was also found that PCS partially mediated the relationship between PI and PC. This is combined with the results showing that 94.6% of the engineers who have had training in management decided to do it for themselves in a manner consistent with Riesman's self-direction [RIE 50, RIE 54] and that 76.8% of them did it to develop competences for a new job as either an internal or an external career move. Of the respondents, 72.7% perceived engineering as a negative status when compared to that of management. This supports the extension of the Schneider [SCH 83, SCH 87, SCH 90] ASA theory from personnel selection and choice to the attraction of a career move, following attrition in engineers experiencing that they have knowledge but managers have power. The attraction proposition is also supported in terms of data on salary and social status.

Exploratory proposition 4: this has implications for differing perspectives on boundaryless or Protean careers.

It has been noted earlier that Arnold *et al.* [ARN 08] have observed that Proteus changed shape not by choice but to avoid capture, which may also be an increasing perception, sense or feeling of being captured in a role that an engineer realizes does not fulfill his or her potential since constrained by narrow job-fit rather than the less bounded and constrained higher level management roles. The findings indicate that this may be among the reasons why engineers now, even if not conscious of the precedents, are seeking to regain the earlier role of engineer-managers.

1.10. Strengths and limits

One of the strengths of the sample, which is broad in itself, was that it was near to equally gender balanced. Another strength of the sample was that those of the engineers holding a management position and those who did not were also near to equally balanced. One of its limits was that it was confined to Portuguese engineers. Another limit was that it did not seek to include ethnicity even if Portugal is not a notably multi-ethnic society.

An innovative strength of the method was that the new PCS construct enhanced the understanding of engineers' perceptions of managers and the relationship between PI and PC. One of the limits of the hierarchical regression technique, although this is shared by any regression analysis, is that this only correlates rather than identifies causality. Another limit was that, although the study was gender balanced, it did not allow for gender differences in perceptions, or relate these to changing needs in work-life balance and the degree to which these may influence opting for management training or education in terms of career choice.

However, there were several remarkably strong findings such as that nearly 73% of the engineers felt themselves to be negatively affected in terms of PCS with managers. Also, that when engineers undertook further education in management, this decision in more than 94% of cases was their own rather than that of the management, while nearly 77% of the engineers decided to do so to develop competences for a new job rather than their current job.

1.11. Implications for future research

Future research on the careers of engineers and their choice to qualify as managers could well be longitudinal in the manner recommended by Arnold and Cohen [ARN 08] and be interview-based rather than only questionnaire-based. It also could assess the degree to which there is a scope, as suggested in two of the exploratory propositions, for extending and assessing the Schneider [SCH 83, SCH 87, SCH 90] ASA principle from personnel selection to career choice as well as assessing the relevance of Foucault's [FOU 72, FOU 77] concept of power-knowledge in terms of how engineers have technical knowledge yet managers without expertise in engineering may exercise power in a manner that directly or indirectly encourages engineers to become managers.

1.12. Bibliography

[ALV 01] ALVESSON M., "Knowledge work: ambiguity, image and identity", *Human Relations*, vol. 54, no. 7, pp. 863–886, 2001.

[ARN 01] ARNOLD J., "Careers and career management", in ANDERSON N., ONES D.S., SINHANGIL H.K., WUSWEVARAN C. (eds), *Handbook of Industrial, Work and Organizational Psychology*, Sage Publications Inc., Thousand Oaks, CA, vol. 2, pp. 115–132, 2001.

[ARN 06] ARNOLD J., LOAN-CLARKE J. *et al.*, "How well can the theory of planned behavior account for occupational intentions", *Journal of Vocational Behavior*, vol. 69, no. 1, pp. 374–390, 2006.

[ARN 08] ARNOLD J., COHEN L., "The psychology of careers in industrial and organizational settings: a critical but appreciative analysis", *International Review of Industrial and Organizational Psychology*, vol. 23, pp. 1–44, 2008.

[ART 94] ARTHUR M.B., "The boundaryless career: a new perspective for organizational inquiry", *Journal of Organizational Behavior*, vol. 15, no. 4, pp. 295–306, 1994.

[ART 95] ARTHUR M.B., CLAMAN P.H. *et al.*, "Intelligent enterprise, intelligent careers", *Academy of Management Executive*, vol. 9, no. 4, pp. 7–20, 1995.

[ART 96] ARTHUR M.B., ROUSSEAU D.M., *The Boundaryless Career – A New Employment Principle for a New Organizational Era*, Oxford University Press, New York, 1996.

[ART 99] ARTHUR M.B., INKSON K. *et al.*, *The New Careers: Individual Action and Economic Change*, Sage, London, 1999.

[ART 05] ARTHUR M.B., KHAPOVA S.N. *et al.*, "Career success in a boundaryless career world", *Journal of Organizational Behavior*, vol. 26, no. 2, pp. 177–202, 2005.

[ASH 89] ASHFORTH B.E., MAEL F., "Social identity theory and the organization", *Academy of Management Review*, vol. 14, no. 1, pp. 20–39, 1989.

[ASH 90] ASHFORTH B.E., GIBBS B.W., "The double-edge of organizational legitimation", *Organization Science*, vol. 1, no. 2, pp. 177–194, 1990.

[BAI 89] BAILYN L., "Understanding individual experience at work: comments on the theory and practice of careers", in ARTHUR M.B., HALL D.T., LAWRENCE B.S. (eds), *Handbook of Career Theory*, Cambridge University Press, New York, pp. 477–489, 1989.

[BAK 96] BAKER T., ALDRICH H.E., "Prometheus stretches: building identity and cumulative knowledge in multiemployer career", in ARTHUR, M.B., ROUSSEAU D.M. (eds), *The Boundaryless Career – A New Employment Principle for a New Organizational Era*, Oxford University Press, New York, pp. 132–149, 1996.

[BAR 89] BARLEY S.R., "Careers, identities, and institutions: the legacy of the Chicago School of Sociology", in ARTHUR M., HALL T., LAWRENCE B. (eds), *Handbook of Career Theory*, Press Syndicate of the University of Cambridge, Cambridge, pp. 41–65, 1989.

[BAR 06] BARLEY S.R., "What we know (and mostly don't know) about technical work", in ACKROYD S. (ed.), *The Oxford Handbook of Work and Organization*, Oxford University Press, Oxford, pp. 376–403, 2006.

[BAR 01] BARTEL C.A., "Social comparisons in boundary-spanning work: effects of community outreach on members' organizational identity and identification", *Administrative Science Quarterly*, vol. 46, no. 3, pp. 379–413, 2001.

[BAR 95] BARTLETT F.C., *Remembering: A Study in Experimental and Social Psychology*, Cambridge University Press, Cambridge, 1995.

[BAR 06] BARUCH Y., "Career development in organizations and beyond: balancing traditional and contemporary viewpoints", *Human Resource Management Review*, vol. 16, no. 2, pp. 125–138, 2006.

[BAU 98] BAUGNET L., *L'identité sociale*, Dunod, Paris, 1998.

[BEC 56] BECKER H.S., CARPER J., "The development of identification with an occupation", *The American Journal of Sociology*, vol. 61, no. 4, pp. 289–298, 1956.

[BEE 08] BEECH N., "On the nature of dialogic identity work", *Organization*, vol. 15, no. 1, pp. 51–74, 2008.

[BER 73] BERGER P., LUCKMANN L., *A Construção Social da Realidade*, Vozes, Petrópolis, 1973.

[BER 32] BERLE A., MEANS G., *The Modern Corporation and Private Property*, Harcourt Brace, New York, 1932.

[BLO 68] BLOCH M., *La société féodale*, Albin Michel, Paris, 1968.

[BOW 10] BOWEN P., EDWARDS P., CATTELL *et al.*, "The awareness and practice of value management by South African consulting engineers: preliminary research survey findings", *International Journal of Project Management*, vol. 28, pp. 285–295, 2010.

[BRA 82] BRAUDEL F., *The Wheels of Commerce*, Collins, London, 1982.

[BRO 03] BROCKLEHURST M., "Self and place: a critique of the boundaryless career", *Critical Management Studies Conference*, Lancaster, 2003.

[BUR 62] BURNHAM J., *The Managerial Revolution*, Penguin Books, Harmondsworth, 1962.

[BUR 81] BURKE P.J., REITZES D.C., "The link between identity and role performance", *Social Psychology Quarterly*, vol. 44, pp. 83–92, 1981.

[BUR 98] BURKE P.J., FRANZOI S.L., "Studying situations and identities using experiential sampling methodology", *American Sociologial Review*, vol. 53, no. 4, pp. 559–568, 1998.

[BUR 04] BURKE P.J., "Identities and social structure: the 2003 Cooley-Mead award address", *Social Psychology Quarterly,* vol. 67, no. 1, pp. 5–15, 2004.

[CAB 04] CABRAL-CARDOSO C., "The evolving Portuguese model of HRM", *The International Journal of Human Resource Management*, vol. 15, no. 6, pp. 959–977, 2004.

[CAP 99] CAPPELLI P., *The New Deal at Work: Managing the Market-Driven Workforce*, Harvard Business School Press, Boston, 1999.

[CHA 62] CHANDLER A.D.J., *Strategy and Structure: Chapters in the History of the Industrial Enterprise*, MIT Press, Cambridge, MA, 1962.

[CHA 08] CHANSON H., "The hydraulics of roman aqueducts: what do we know? Why should we learn?" *World Environmental and Water Resources Congress, ASCE-EWRI Education, Research and History Symposium*, Hawaii, 2008.

[CLA 09] CLARKE C.A., BROWN A.D. *et al.*, "Working identities? Antagonistic discursive resources and managerial identity", *Human Relations*, vol. 62, no. 3, pp. 323–352, 2009.

[CLE 05] CLEGG S.R., KORNBERGER M. *et al.*, "Learning/organizing/becoming", *Organization*, vol. 12, no. 2, pp. 147–168, 2005.

[CLE 07] CLEGG S.R., RHODES C. *et al.*, "Desperately seeking legitimacy: organizational identity and emerging industries", *Organizational Studies*, vol. 28, no. 4, pp. 495–513, 2007.

[COL 00a] COLLIN A., YOUNG R.A., "The Future of Career", in COLLIN A., YOUNG R.A. (eds), *The Future of Career*, Cambridge University Press, Cambridge pp. 276–300, 2000.

[COL 00b] COLENSO M., *Kaizen Strategies for Organisational Change: Evolution and Revolution in the Organisation*, Financial Times-Prentice Hall, London, 2000.

[COO 90] COOLEY M., *Architect or Bee?*, Hogarth Press, London, 1990.

[COS 05] COSTELLO C.Y., *Professional Identity Crisis: Race, Class, Gender and Success at Professional Schools*, Vanderbilt University Press, 2005.

[DE 08a] DE VOS A., DEWETTINCK K. *et al.*, "The professional career on the right track: a study on the interaction between career self-management and organizational career management in explaining employee outcomes", *European Journal of Work and Organizational Psychology*, vol. 1, no. 18, pp. 55–80, 2008.

[DE 08b] DE VOS A., SOENS N., "Protean attitude and career success: the mediating role of self-management", *Journal of Vocational Behavior*, vol. 73, no. 1, pp. 449–456, 2008.

[DEF 94] DEFILLIPPI R.J., ARTHUR M.B., "The boundaryless career: a competency-based perspective", *Journal of Organizational Behavior*, vol. 15, no. 4, pp. 307–324, 1994.

[DIC 04] DICK R.V., WAGNER U. *et al.*, "The utility of a broader conceptualization of organizational identification: which aspects really matter?", *Journal of Occupational & Organizational Psychology*, vol. 77, no. 2, pp. 171–191, 2004.

[DIC 05] DICK R.V., WAGNER U. *et al.*, "Category salience and organizational identification", *Journal of Occupational & Organizational Psychology*, vol. 78, no. 1, pp. 273–285, 2005.

[DIC 06] DICK P., HYDE R., "Line manager involvement in work-life balance and career development: can't manage, won't manage?", *British Journal of Guidance & Counselling*, vol. 34, no. 3, pp. 345–364, 2006.

[DIM 83] DIMAGGIO P.J., POWELL W.W., "The iron cage revisited: institutional isomorphism and collective rationality in organizational fields", *American Sociological Review*, vol. 48, no. 2, pp. 147–160, 1983.

[DOW 09] DOWN S., REVELEY J., "Between narration and interaction: situating first-line supervisor identity work", *Human Relations*, vol. 62, no. 3, pp. 379–401, 2009.

[DOY 00] DOYLE M., "Managing careers in organizations", in COLLIN A., YOUNG R.A. (eds), *The Future of Career*, Cambridge University Press, Cambridge, pp. 228–242, 2000.

[DUB 96] DUBAR C., *La socialisation: construction des identités sociales et professionnelles*, Armand Colin, Paris, 1996.

[DUB 97] DUBAR C., *A Socialização – construção de identidades sociais e profissionais*, Porto Editora, Porto, 1997.

[DUN 98] DUNLAP W.P., LANDIS, R.S., "Interpretations of multiple regression borrowed from factor analysis and canonical correlation", *The Journal of General Psychology*, vol. 4, no. 125, pp. 397–407, 1998.

[DUR 57] DURKHEIM E., *Professional Ethics and Civic Morals*, Routledge & Kegan Paul, London, 1957.

[DUT 94] DUTTON J.E., DUKERICH J.M. *et al.*, "Organizational images and member identification", *Administrative Science Quarterly*, vol. 39, no. 2, pp. 239–263, 1994.

[ELS 01] ELSBACH K.D., BHATTACHARYA C.B., "Defining who you are by what you are not: organizational disidentification and the National Rifle Association", *Organization Science*, vol. 12, no. 4, pp. 393–403, 2001.

[EPS 90] EPSTEIN S., "Cognitive-experiential self-theory", in PERVIN L. (ed.), *Handbook of Personality: Theory and Research*, Guilford Press, New York, 1990.

[EPS 91] EPSTEIN S., "Cognitive-experiential self-theory: implications for developmental psychology", in GUNNAR M.R., SROUFE L.A. (eds), *Self-Process and Development*, vol. 23, Erlbaum, Hillsdale, NJ, pp. 79–123, 1991.

[EPS 93] EPSTEIN S., "Emotion and self-theory", in LEWIS M.A., HAVILAND J. (eds), *Handbook of Emotions*, Guilford Press, New York, 1993.

[EPS 94] EPSTEIN S., "Integration of the cognitive and psychodynamic unconscious", *American Psychologist*, vol. 49, no. 8, pp. 709–724, 1994.

[FOU 72] FOUCAULT M., *The Archaeology of Knowledge*, Pantheon Books, New York, 1972.

[FOU 77] FOUCAULT M., *Discipline and Punish*, Allen Lane, London, 1977.

[FOU 99] FOURNIER V., "The appeal to professionalism as a disciplinary mechanism", *Sociological Review*, vol. 47, no. 22, pp. 280–307, 1999.

[GIO 98] GIOIA D.A., "From individual to organizational identity", in WHETTEN D.A., GODFREY P.C. (eds), *Identity in Organizations: Building Theory Through Conversations*, Sage Publications, Inc., Thousand Oaks, CA, pp. 17–31, 1998.

[HAL 76] HALL D.T., *Careers in Organizations*, Scott Foresman Co., Glenview, 1976.

[HAL 02] HALL D.T., *Careers In and Out of Organizations*, Sage, Thousand Oaks, CA, 2002.

[HAL 04] HALL D.T., "The protean career: a quarter-century journey", *Journal of Vocational Behaviour*, vol. 65, no. 1, pp. 1–13, 2004.

[HAL 05] HALL D.T., CHANDLER D.E., "Psychological success: when the career is a calling", *Journal of Organizational Behavior*, vol. 26, no. 2, pp. 155–176, 2005.

[HAR 05] HARRIS G., CAMERON J., "Multiple dimensions of organizational identification and commitment as predictors of turnover intentions and psychological well being", *Canadian Journal of Behavioral Science*, vol. 37, no. 3, pp. 159–169, 2005.

[HAV 94] HAVILAND J.M., DAVIDSON R.B. *et al.*, "The place of emotion in identity", *Journal of Research on Adolescence*, vol. 4, no. 4, pp. 503–518, 1994.

[HEK 09] HEKMAN D.R., "Combined effects of organizational and professional identification on the reciprocity dynamic for professional employees", *Academy of Management Journal*, vol. 52, no. 3, pp. 506–526, 2009.

[HES 05a] HESLIN P.A., "Conceptualizing and evaluating career success", *Journal of Organizational Behavior*, vol. 26, no. 2, pp. 113–136, 2005.

[HES 05b] HESLIN P.A., "Experiencing career success", *Organizational Dynamics*, vol. 34, no. 4, pp. 376–390, 2005.

[HIL 06] HILDEBRANDT E., LITTIG B., "Concepts, approaches and problems of work-life balance", *European Societies*, vol. 8, no. 2, pp. 215–222, 2006.

[HIN 89] HINKLE S., TAYLOR L.A., *et al.*, "Intragroup identification and intergroup differentiation: a multicomponent approach", *British Journal of Social Psychology*, vol. 28, no. 4, pp. 305–317, 1989.

[HIR 70] HIRSCHMAN A., *Exit, Voice and Loyalty*, Harvard University Press, Cambridge, MA, 1970.

[HOD 01] HODGE A.T., *Roman Aqueducts & Water Supply*, Duckworth, London, 2001.

[HOG 00] HOGG M.A., TERRY D.J., "Social identity and self-categorization processes in organizational contexts", *Academy of Management Review*, vol. 25, no. 1, pp. 121–140, 2000.

[HOL 13] HOLLAND S., OLIVEIRA T.C., "Who misplaced post modernism? Hume, Adam Smith and economic methodology", *Economic Theory*, forthcoming.

[HOW 00] HOWARD J., "The social psychology of identities", *Annual Review of Sociology*, vol. 26, no. 1, pp. 367–393, 2000.

[HUM 02] HUMPHREYS M., BROWN A.D., "Narratives of organizational identity and identification: a case study of hegemony and resistance", *Organization Studies*, vol. 23, no. 3, pp. 421–447, 2002.

[INK 04] INKSON K., "Images of career: nine key metaphors", *Journal of Vocational Behavior*, vol. 65, no. 1, pp. 96–111, 2004.

[INK 07] INKSON K., *Understanding Careers: The Metaphors of Working Lives*, Sage Publications, Inc, Thousand Oaks, CA, 2007.

[JON 07] JONES C., DUNN M.B., "Careers and institutions: the centrality of careers to organizational studies", GUNZ H., PEIPERL M. (eds), *Handbook of Career Studies*, Sage Publications, Thousand Oaks, CA, 2007.

[JUD 04] JUDGE T.A., KAMMEYER-MUELLER J., *et al.*, "A longitudinal model of sponsorship and career success: a study of industrial-organizational psychologists", *Personnel Psychology*, vol. 57, no. 2, pp. 271–303, 2004.

[KAN 89] KANTER R.M., *When Giants Learn to Dance: Mastering the Challenges of Strategy, Management and Careers in the 1990's London*, Unwin, 1989.

[KEP 91] KEPPEL G., *Design and Analysis: A Researcher's Handbook*, Prentice-Hall, Englewood Cliffs, NJ, 1991.

[KIN 04] KING Z., "Career self-management: its nature, causes and consequences", *Journal of Vocational Behavior*, vol. 65, no. 1, pp. 112–133, 2004.

[KIN 05] KING Z., BURKE S. *et al.* "The 'bounded' career: an empirical study of human capital, career mobility and employment outcomes in a mediated labour market", *Human Relations*, vol. 58, no. 8, pp. 981–1007, 2005.

[KIR 06] KIRCHMEYER C., "The different effects of family on objective career success across gender: a test of alternative explanations", *Journal of Vocational Behavior*, vol. 68, pp. 323–346, 2006.

[KNI 00] KNIPPENBERG D.V., SCHIE E.C.M.V., "Foci and correlates of organizational identification", *Journal of Occupational & Organizational Psychology*, vol. 73, no. 2, pp. 137–147, 2000.

[KRO 76] KRONUS C.L., "Occupational versus organizational influences on reference group identification: the case of pharmacy", *Sociology of Work and Occupations*, vol. 3, no. 3, pp. 303–330, 1976.

[LAC 87] LACEY R., *Ford: The Men and the Machine*, Heinemann and Pan Books, London, 1987.

[LEE 92] LEE G.L., SMITH C., (ed.), *Engineers and Management: International Comparisons*, Routledge, New York, 1992.

[LEE 06] LEE C.-H., HSU M.-L. *et al.*, "The impacts of benefit plans on employee turnover: a firm-level analysis approach on Taiwanese manufacturing industry", *The International Journal of Human Resource Management*, vol. 17, no. 11, pp. 1951–1975, 2006.

[LEE 10] LEE J.J., KANG S.Y., Heo J.-H., *Business and Management Competency of Engineers*, IEEE EDUCON – Education Engineering 2010 – The Future of Global Learning Engineering Education, Madrid, Spain, 2010.

[LIP 90] LIPIANSKY E.M., "Introduction à la problématique de l'identité", in CAMILLERI C., LIPIANSKY E.-M., KASTERSZTEIN J., MALEWSKA-PEYRE H. (eds), *Stratégies Identitaires*, Presses universitaires de France, 2002, pp. 7–26, 1990.

[LIT 03] LITTLER C.R., WIESNER R. *et al.*, "The dynamics of delayering: changing management structures in three countries", *Journal of Management Studies*, vol. 40, no. 2, pp. 225–256, 2003.

[LOK 10] LOK J., "Institutional logics as identity projects", *Academy of Management Journal*, vol. 53, no. 6, pp. 1305–1335, 2010.

[LON 93] LONDON M., "Relationships between career motivation, empowerment and support for career development", *Journal of Occupational & Organizational Psychology*, vol. 66, no. 1, pp. 55–69, 1993.

[MAC 03] MACHADO H.V., "A identidade e o contexto organizacional: perspectivas de análise", *Revista de Administração Contemporânea*, vol. 7 (Edição Especial), pp. 51–73, 2003.

[MAS 43] MASLOW A.H., "A theory of human motivation", *Psychology Review*, vol. 50, pp. 370–396, 1943.

[MAT 88] MATTE BLANCO I., *Thinking, Feeling and Being*, Routledge, London, 1988.

[MEY 93] MEYER J.P., ALLEN N.J., SMITH C.A., "Commitment to organizations and occupations: extension and test of a three component conceptualization", *Journal of Applied Psychology*, vol. 4, no. 78, pp. 538–551, 1993.

[MIG 09] MIGUEL S., "Desempenho profissional numa Organização de Saúde: Um modelo de análise", *Revista Portuguesa e Brasileira de Gestão*, vol. 8, no. 4, pp. 37–53, 2009.

[MIN 94] MINTZBERG H., *The Rise and Fall of Strategic Planning*, The Free Press, New York, 1994.

[MIR 94] MIRVIS P.H., HALL D.T., "Psychological success and the boundaryless career", *Journal of Organizational Behavior*, vol. 15, no. 4, pp. 365–380, 1994.

[MIT 99] MITHEN S., *The Prehistory of the Mind: The Cognitive Origins of Art, Religion and Science*, Thames and Hudson, London, 1999.

[NEE 86] NEEDHAM J., *Science and Civilization in China*, Caves Books, Taipei, 1986.

[NG 05] NG T.W.H., EBY L.T. *et al.*, "Predictors of objective and subjective career success: a meta-analysis", *Personnel Psychology*, vol. 58, no. 2, pp. 367–408, 2005.

[NG 07] NG T.W.H., SORENSEN K.L. *et al.*, "Determinants of job mobility: a theoretical integration and extension", *Journal of Occupational & Organizational Psychology*, vol. 80, no. 3, pp. 363–386, 2007.

[NIC 05] NICHOLSON N., DE WAAL-ANDREWS W., "Playing to win: biological imperatives, self-regulation, and trade-offs in the game of career success", *Journal of Organizational Behavior*, vol. 26, no. 2, pp. 137–154, 2005.

[OLI 00] OLIVEIRA T.C., "Implicit logic in unstructured interviewing, selection & development review", *British Psychological Society*, vol. 16, no. 2, pp. 10–14, 2000.

[OLI 06] OLIVEIRA T.C., "Implicit logic in managerial decision-making", *Notas Económicas*, Faculty of Economics, University of Coimbra, 2006.

[OLI 07a] OLIVEIRA T.C., "Delving down to learn up", *Plenary Presentation to the Sloan MIT-Portugal Conference on New Developments in Management*, Lisbon, 2007.

[OLI 07b] OLIVEIRA T.C., HOLLAND S., "Alternative paradigms of hospital work organisation and health provision", *Revista Portuguesa de Saúde Pública:*, vol. 25, no. 1, pp. 19–37, 2007.

[OLI 12] OLIVEIRA T.C., HOLLAND S., "On the centrality of human value", *Journal of Economic Methodology*, vol. 19, no. 2, pp. 121–141, 2012.

[OLI 13] OLIVEIRA T.C., HOLLAND S., "Institutional logics enabling and inhibiting innovation", in MACHADO C., DAVIM J.P. (eds), *Management and Engineering Innovation*, ISTE Ltd., London, Wiley, New York, 2013.

[PER 07] PEREIRA H., Os enfermeiros e a identidade profissional, 2007. Available at http://www.ordemenfermeiros.pt/index.print.php?page=72&view=news:Print &id=261.

[POP 59] POPPER K., *The Logic of Scientific Discovery*, Hutchinson, London, 1959.

[RAA 07] RAABE B., FRESE M. *et al.*, "Action regulation theory and career self-management", *Journal of Vocational Behavior*, vol. 70, no. 1, pp. 297–311, 2007.

[RIE 50] RIESMAN D., GLAZER N., DENEY R., *The Lonely Crowd*, Yale University Press, 1950.

[RIE 54] RIESMAN D., *Individualism Reconsidered and Other Essays*, The Free Press, Glencoe, IL, 1954.

[ROB 06] ROBSON S.M., HANSSON R.O. *et al.*, "Successful aging: criteria for aging well in the workplace", *Journal of Career Development*, vol. 33, no. 2, pp. 156–177, 2006.

[ROT 00] ROTHSCHILD E., *Economic Sentiments: Adam Smith, Condorcet and the Enlightenment*, Harvard University Press, Cambridge MA, 2000.

[ROU 95] ROUSSEAU D.M., *Psychological Contracts in Organizations: Understanding Written and Unwritten Agreements*, Sage, Newbury Park, CA, 1995.

[ROU 98] ROUSSEAU D.M., "The problem of the psychological contract", *Journal of Organisational Behavior*, vol. 19, pp. 665–671, 1998.

[RYN 04] RYNES S.L., GERHART B. *et al.*, "The importance of pay in employee motivation: discrepancies between what people say and what they do", *Human Resource Management*, vol. 43, no. 4, pp. 381–394, 2004.

[SAR 57] SARTRE J.-P., *L'existentialisme est un humanisme*, Nagel, Paris, 1957.

[SCH 18] SCHOPENHAUER A., *The World as Will and Representation*, 1st English translation, 1883, 3 vols, Routledge and Kegan Paul, London, see also (1969), Dover Publications, New York, 2 vols, 1818.

[SCH 49] SCHUMPETER J., *The Theory of Economic Development*, Harvard University Press, Cambridge, MA, 1949.

[SCH 83] SCHNEIDER B., "Interactional psychology and organisation behaviour", in STAW B.M., CUMMINGS L.L. (eds), *Research in Organisation Behaviour*, vol. 5, JAI Press, Greenwich City, pp. 1–31, 1983.

[SCH 87] SCHNEIDER B., "The people make the place", *Personnel Psychology*, vol. 40, pp. 437–453, 1987.

[SCH 90] SCHNEIDER B., *Organizational Climate and Culture*, Jossey-Bass, San Francisco, CA, 1990.

[SCH 96] SCHEIN E.H., "Career anchors revisited: implications for career development in the 21st century", *Academy of Management Executive*, vol. 10, no. 4, pp. 80–88, 1996.

[SCO 01] SCOTT W.R., *Institutions and Organizations*, Sage Publications, Inc, Thousand Oaks, CA, 2001.

[SEN 90] SENGE P., *The Fifth Discipline*, Doubleday, New York, 1990.

[SHA 98] SHAW V.S., CHRISTOPHER T., "Conflict between engineers and marketeers: the engineer's perspective", *Industrial Marketing Management*, vol. 22, pp. 299–309, 1998.

[SHA 03] SHAW V.S., CHRISTOPHER T., ENKE M., "Conflict between engineers and marketeers: the experience of German engineers", *Industrial Marketing Management*, vol. 32, pp. 489–499, 2003.

[SIM 05a] SIMS D., "Living a story and storying a life: a narrative understanding of distributed self", in LINSTEAD A., LINSTEAD S. (eds), *Organization and Identity*, Routledge, London, pp. 86–104, 2005.

[SIM 05b] SIMS D., "You bastard: a narrative exploration of the experience of indignation within organizations", *Organization Studies*, vol. 26, no. 11, pp. 1625–1640, 2005.

[SLO 64] SLOAN A., *My Years with General Motors*, Doubleday, New York, 1964.

[SMI 76] SMITH A., *An Enquiry into the Nature and Causes of the Wealth of Nations*, Dent, London, 1776.

[STU 08] STURGES J., CONWAY N. *et al.*, "What's the deal? An exploration of career management behaviour in Iceland", *The International Journal of Human Resource Management*, vol. 19, no. 4, pp. 752–768, 2008.

[TAJ 79] TAJFEL H., TURNER J.C., "An integrative theory of intergroup conflict", in WORCHEL S., AUSTIN W.G. (eds), *The Social Psychology of Intergroup Relations*, Brooks Cole Publishing Company, Monterey, pp. 33–47, 1979.

[TAJ 86] TAJFEL H., TURNER J.C., "The social identity theory of intergroup behaviour", in WORCHEL S., AUSTIN W.G. (eds), *Psychology of Intergroup Relations*, Nelson-Hall, Chicago, pp. 7–24, 1986.

[THA 99] THARENOU P., "Is there a link between family structures and women's and men's managerial career advancement?", *Journal of Organizational Behavior*, vol. 20, no. 6, pp. 837–862, 1999.

[THA 08] THATCHER S.M.B., GREER L.L., "Does it really matter if you recognize who I am? The implications of identity comprehension for individuals in work teams", *Journal of Management*, vol. 34, no. 1, pp. 5–24, 2008.

[TUR 87] TURNER J.C., HOGG M. *et al.*, *Rediscovering the Social Group: Self-Categorization Theory*, B. Blackwell, Oxford, 1987.

[VAN 77a] VAN MAANEN J., *Experiencing Organization: Notes on the Meaning of Careers and Socialization*, Wiley, New York, 1977.

[VAN 77b] VAN MAANEN J., SCHEIN E.H., "Career development", in HACKMAN J.R., SUTTLE J.L. (eds), *Improving Life at Work*, Goodyear Publishing, Santa Monica, CA, pp. 30–95, 1977.

[VAR 12] VAROUFAKIS Y., HOLLAND S., "A modest proposal for resolving the eurozone crisis", *Intereconomics*, vol. 47, no. 4, pp. 240–247, 2012.

[VAU 91] VAUGHAN A., *Isambard Kingdom Brunel: Engineering Knight-Errant*, John Murray, London, 1991.

[VEB 21] VEBLEN T., *The Engineers and the Price System*, Kitchener Ontario Batoche Books, 1921.

[VER 66] VERNON R., "International investment and international trade in the product cycle", *Quarterly Journal of Economics*, vol. 80, no. 2, pp. 190–207, 1966.

[WOM 96] WOMACK J.P., JONES D.T., *Lean Thinking*, Simon and Schuster, New York, 1996.

[WOM 05] WOMACK J.P., JONES D.T., *Lean Solutions: How Companies and Customers Can Create Wealth Together*, Simon and Schuster, New York, 2005.

[YOU 92] YOUNG R.A., COLLIN A., *Interpreting Career. Hermeneutical Studies of Lives in Context*, Praeger, Westpoint, London, 1992.

[YOU 04] YOUNG R.A., COLLIN A., "Constructivism and social constructionism in career field", *Journal of Vocational Behavior*, vol. 64, no. 3, pp. 373–398, 2004.

[ZUC 77] ZUCKER L.G., "The role of institutionalization in cultural persistence", *American Sociological Review*, vol. 42, no. 1, pp. 726–743, 1977.

Chapter 2

Strategic Capabilities for Successful Engagement in Proactive CSR in Small and Medium Enterprises: A Resource-Based View Approach

The objectives of this chapter are twofold: first, to review and discuss the literature on how the distinctive organizational characteristics attributed to small and medium enterprises (SMEs) can provide the opportunity to gain a strategic advantage through the development of specific capabilities for "proactive" corporate social responsibility (CSR); and second to propose a model explaining how SMEs can generate positive outcomes through the adoption of business practices based on such strategic capabilities, that is improve financial performance while enabling progress toward meeting the call by society for responsible business strategies.

The chapter contains a discussion of six particular strategic capabilities, associated with SMEs in the literature and which our research suggests make the adoption of proactive CSR more likely. These specific capabilities comprise three "internally directed" (shared vision, organizational learning and capital management) and three "externally directed" (stakeholder engagement, strategic proactivity and continuous innovation). We argue that when an SME

Chapter written by Nuttaneeya (Ann) TORUGSA and Wayne O'DONOHUE.

engages in a strategy that efficiently exploits and enhances these capabilities, competitive advantage and superior performance will be achieved.

2.1. Introduction – CSR and SMEs

2.1.1. *The nature of CSR*

CSR and its role in the generation of competitive advantage are now highly visible topics on many research and managerial agendas [GRE 04, MCW 06]. Given impetus by the argument that "not only is doing good the right thing to do, but it also leads to doing better" [BHA 04, DUN 03, KOT 05], various theoretical approaches aimed at understanding how firms can maximize financial returns while enabling progress toward meeting the call for responsible business strategies have been developed, thus contributing to significant advances in knowledge about the effective implementation of CSR [LIN 10]. However, despite these developments, we suggest that currently there is no definition of CSR that is uniformly accepted, a situation which reflects in part its broad conceptual nature and consequently its overlap with, and similarity to, other conceptions of business–society relations [MAT 08].

We draw on the notion of CSR (also called "responsible entrepreneurship") offered by the European Commission [EUR 03] to define CSR as responsible business strategies and practices which support and integrate the three principles of sustainability: economic growth and prosperity, social cohesion and equity, and environmental integrity and protection, and which thereby produce an overall positive impact on the firm and society. We also assume that every firm has a societal responsibility but expect, in-line with the long-established "reaction-defence-accommodation-proaction" typology [CAR 79], that the degree and kind of CSR strategies used by a firm to meet its responsibility will be contingent on a range of factors [WAR 85, WIL 75]. In this light, firms can be viewed as operating along a continuum of CSR ranging from "reactive" to "proactive" in nature.

Firms engaging in reactive CSR expend the minimum level of effort required for non-voluntary regulatory compliance [CAR 79, GRO 11, WIL 75]. Such firms meet minimum industry norms and standards and comply with legislation requiring attention simply to gain and maintain social legitimacy (i.e. a firm's "licence to operate"); CSR is not seen as a source of competitive advantage and financial benefits that warrants discretionary expenditure. In contrast, firms engaging in proactive CSR often attempt to take a leading position on that basis, even at times when institutional pressures for such action have not yet taken hold. To achieve leadership, such firms voluntarily adopt strategies and practices that go beyond regulatory compliance in order to manage CSR issues as a competitive priority [CAR 79, DU 07, GRO 11, WIL 75]. The adoption of proactive CSR that embraces the design and development of sustainable products, production process and operations to anticipate external regulation and social trends, is commonly advocated by strategy scholars as a value-creating action from which a competitive advantage and superior performance can be derived [GRO 11, KLA 99, SHA 98]. It is this proactive approach to CSR upon which we focus in our discussion here.

2.1.2. *The SME context*[1]

While the nature and implementation of proactive CSR has been well researched in large enterprises, SMEs have so far received less research attention. Given the shortage of publicly available data and the lesser degree of empirical research evidence pertaining to SMEs [RUT 00], it is widely presumed that legislative pressures produce only a reactive approach to CSR among SMEs. This presumption may reflect a general view that SMEs must, of necessity, focus primarily on sustaining financial viability, and will therefore invest their limited resources in ways that yield an immediate financial return. As a result, once compliance requirements are met, SMEs are left with no or inadequate resources to engage proactively in CSR [RUS 97, SCH 02, TIL 99]. The generalizability of this view is however questioned by some researchers [FIT 11, RUS 09], who argue that proactive CSR may indeed be occurring in SMEs, but because the language of CSR is not often used to describe such activity, it goes unrecognized as such.

---

1 For the purposes of this discussion, we have deliberately chosen not to wander unaccompanied into the murky shadows that obscure the question of how SME size should be measured. While staffing, financial or other measures can be used, the appropriate choice will depend in part on context. The specific research providing the impetus for this chapter was done in the Australian business context and used staff numbers as the demarcation measure. In the Australian economy, SMEs are commonly defined as those firms with fewer than 200 employees by the Australian Bureau of Statistics [AUS 01]; not surprisingly, this scale of firm size is generally well below that of firms in larger economies, such as the United States, where firms with fewer than 500 employees are usually categorized as "small" [BAR 07]. However, we do not believe the lack of either a universal scale for defining an SME or a common economic context need form a barrier to the reader's understanding of what is covered in this chapter.

Nevertheless, limited resources and expertise, as well as die-hard perceptions that implementing proactive CSR is antagonistic to maximizing profits, do form significant stumbling blocks to the implementation of sustainability principles. In such light, a well-thought-out strategic approach to proactive CSR is therefore essential if SMEs are to cope with the complex challenges they face and thus generate wealth in a manner that is societally responsible. Assisting SMEs in this process has arguably become a key challenge for researchers, a challenge acknowledged by the call in the literature for a reorientation of the research focus on proactive CSR away from large enterprises and toward SMEs (see [ARA 08, JEN 09, LIN 10, TOR 12a, TOR 12b]). The underlying reasons for this call are threefold.

First, SMEs are a significant part of every economy and, in most industrialized nations, are accepted as being the predominant employers, innovation drivers and wealth generators [HAW 06]. As reported by the United Nations Industrial Development Organization, SMEs make up over 90% of business worldwide and account for around 60% of employment [UNI 02]. By emphasizing labor-intensive production processes, SMEs contribute significantly to the provision of productive employment opportunities, the generation of income and ultimately the reduction of poverty. In this way, SME prosperity is inextricably linked to a nation's economic and social health [UNI 02, TIL 00]. Second, the cumulative impact of SMEs on the environment and society, when viewed from a sustainability perspective, raises concerns. Research suggests that in global terms SMEs produce over 60% of global carbon dioxide emissions from business [MAR 98, SMI 98], and their combined environmental impacts outweigh those of large firms [HIL 00]. It is, however, interesting to note, based on the finding of a recent survey of 800 Australian SMEs across 14 industry sectors conducted by the Australian Fieldwork Solution [AUS 09], that over 70% of respondents indicated their business operations do not harm the environment. Such a perception may perhaps explain in part the generally low commitment to making environmental improvements that the survey revealed. Finally, there is consensus in the strategic management literature that the strategic distinctiveness and organizational characteristics of SMEs militate against assuming "small firms" are really only "little large firms" [ARA 08, BAR 01, DEA 98, TOR 12a, TOR 12b]. In other words, the strategic differences between SMEs and large firms, the significant impacts of SMEs on economy, society and the environment, as well as the steep learning curve SMEs must master to seize new opportunities, survive and indeed prosper in the now highly competitive global business environment, all support the need for giving greater attention to the relationship between firm size and strategic advantage with regard to proactive CSR.

Initially, there was a widely held view in the literature [GAL 72, STR 77] that large enterprises possess numerous advantages over SMEs that enable growth-oriented strategies designed to capture economies of scale, brand recognition and market power. In contrast, SMEs were generally considered as being disadvantaged by their lesser and limited resource base, constrained organizational learning opportunities, restricted production capacity and narrower range of markets [GIB 85, STA 76]. Moreover, SMEs were often marginalized in the literature as a residual class of enterprise that had "failed" to become large, and were less likely to use advanced managerial approaches, strategic analysis and planning practices [ROB 84, SEX 85, SHU 85]. In recent times, however, the strategic management literature has reflected increasing interest in the proposition that, in comparison to their larger counterparts, SMEs are better able to develop particular strategic capabilities that provide a distinct basis for strategic advantage based on organizational characteristics, such as: simple management structures, more flexibility, shorter lines of communication, closer employer–employee interaction, coincidence of ownership and control, greater responsiveness to change in the business environment and stakeholder needs, better entrepreneurial alertness, creativeness and innovativeness, and risk-taking behaviors [FIE 91, MAN 02, RAN 99, SMA 00, VOS 98, YU 01]. These characteristics have been identified in recent empirical studies [ARA 08, TOR 12a, TOR 12b] as enabling SMEs to develop particular organizational capabilities that contribute substantially to competitive advantage built on proactive CSR principles. Such research suggests that while resource constraints can impose barriers to SMEs making significant investments in proactive CSR, they may not be the determining factor, and thus proactive CSR, when adopted and implemented strategically, can be a pathway to competitive advantage for SMEs.

2.2. The resource-based view approach to business strategy

2.2.1. *Resources and capabilities*

Generally, research in business strategy has usually aimed at developing and testing theoretical models or frameworks which firms can apply in choosing strategies that contribute to competitive advantage and generate superior performance. Much of this research has been informed by a resource-based view (RBV) of the firm. The RBV approach has two basic assumptions: first, it assumes that a firm's resources and capabilities may allow it to implement strategies that alter structural aspects of its industry in ways that uniquely benefit that firm; and second it assumes that resources and capabilities are heterogeneously distributed and imperfectly mobile among firms in an industry [BAR 91, PEN 59, RUM 84, WER 84]. These two fundamental assumptions allow not only for the existence of differences between firm resource endowments, but also for such differences to persist over time. For these reasons, firms should choose strategies that most completely exploit their resource and capability distinctiveness to advantage themselves over their competitors. By building a competitive advantage in this way, a firm may generate superior returns.

The RBV emphasis on distinctive resources and capabilities does not mean that the industry structure and external environment are not important; on the contrary, it allows that the choice of which attributes firms should nurture and exploit, must be guided by an understanding of industry structure, the positional advantages being sought and trends in society and environment [BAR 86a, CON 91, DAY 94, GRA 91]. As Wernerfelt [WER 84, p. 171] puts it, "for the firm, resources and products are two sides of the same coin", meaning that distinctiveness in the product offering is tied to distinctiveness in the resources and capabilities used to produce the product. Arguably, since the RBV approach views the resources and capabilities of a firm as the primary foundation upon which a firm can establish its identity and frame its strategy and, as such, the primary sources of a firm's competitive advantage and superior performance [BAR 91, GRA 91, WER 84], the key to understanding a resource-based approach to strategy formulation is to understand the relationships between resources, capabilities, strategies, competitive advantage and performance, as well as the mechanisms through which competitive advantage can be sustained over time.

RBV theory distinguishes resources and capabilities generally in the following ways. Resources, on the one hand, are considered the stocks of available tangible factors (e.g. financial resources, human resources and physical capital resources, including plant, location, equipment and raw materials) and intangible factors (e.g. intellectual property rights, know-how of personnel, informant networks, organizational culture and reputation), which are owned or controlled by the firm and which enable it to produce efficiently [BEC 64, WIL 75]. Capabilities, on the other hand, are considered to derive from the firm's capacity to bring its resources together and deploy them advantageously [BAR 91, DAY 94]. In essence, it is the way a firm integrates and uses its resources to create capabilities that form the main source of its competitive advantage and performance improvement [AMI 93, GRA 91, PET 93].

Capabilities can vary in terms of the degree to which they have an internal or external focus [GRA 98, SHA 07]. Some capabilities which we consider as "internally directed" are mainly embedded in routines and processes that generate and diffuse knowledge and learning within the firm in order to increase efficiencies, reduce costs and so facilitate the firm's viability over the long term. Other capabilities that are "externally directed" consist more of routines and processes that are focused toward exploiting opportunities in the external business or social context within which a firm operates, in order to maintain a strategic fit with that environment. This broad two-part classification system is similar to the "inside-out and outside-in capabilities" [MOO 99], "fan-out and fan-in capabilities" [HAR 04] and "internally-focused and externally-focused capabilities" [SHA 07], discussed in the literature.

While all capabilities can be so classified, under an RBV approach not all capabilities possessed by a firm are considered to be of equal significance and therefore worthy of equal attention; rather, the primary focus of attention is on those capabilities which are distinctive and thus core to a firm's sustained competitive advantage. To determine such core capabilities, various scholars have proposed a number of criteria [AMI 93, BAR 91, COL 98, GRA 91, HAL 93]. Rangone [RAN 99, p. 234] has summarized the most important criteria in terms of: *competitive superiority* (the extent to which a capability contributes to differentiating the firm from its rivals), *imitability* (the degree of difficulty actual and potential competitors face in imitating the capability because of its uniqueness, path dependency or causal ambiguity), *duration* (or long-term benefits of the capability), *appropriability* (the extent to which the firm is able to appropriate the advantages and market opportunities a capability generates) and *substitutability* (the extent to which rival firms can substitute the capability with an alternative that gives the same advantages). These criteria are considered to be appropriate regardless of the firm size.

Various RBV-based frameworks have been developed to support strategy analysis and planning processes [AMI 93, COL 98, HAL 93, PET 93, PRA 90]. One widely accepted framework is that developed by Grant [GRA 91], which consists of five stages: (1) identifying the firm's resource base; (2) appraising the firm's capabilities relative to competitors; (3) analyzing the profit-earning potential of the firm's resources and capabilities; (4) selecting a strategy that best exploits the firm's resources and capabilities and that fits to the opportunities available in the competitive environment; and (5) extending and supporting the development of new distinctive resources and capabilities arising out of strategy implementation. On the basis of this approach, competitive advantage and superior performance result from a strategy of planned enhancement and exploitation of a firm's resources and capabilities. This approach is also considered to be appropriate regardless of the firm size.

2.2.2. *Competitive advantage and performance*

Turning to the terms "competitive advantage" and "performance", although these two constructs are often used interchangeably [POR 85], they are in fact acknowledged to be distinct [POW 01]. Competitive advantage is generally conceptualized as resulting from the implementation of a strategy not currently being implemented by competitors, and which facilitates the reduction of costs and/or exploitation of market opportunities through product differentiation [BAR 91, POR 85]. Performance is generally conceptualized in financial terms as what a firm earns as a result of the implementation of its strategies [RUM 94]. Barney [BAR 97, p. 63], who defines performance in terms of "comparing actual value created by a firm with its expected value", distinguishes between three levels of performance, specifically between firms that earn: (1) just what is expected by owners or a "normal" level of performance; (2) less than expected or a "below-normal" level of performance; and (3) more than expected or a "superior above-normal" level of performance. As a point to note, linking implementation of strategy to competitive advantage in this way does not necessarily imply that strategy is always a sufficient condition for improved financial performance [COF 99, RUM 91, PET 03]. Even when a firm does effectively implement a strategy, it may often find that it is unable to recoup the resulting financial value in full at a cost lower than that required to create it. Indeed, given that many factors extraneous to the firm may negatively affect performance, a firm's financial performance may reduce even in the existence of a well-executed strategy. Therefore, to achieve and sustain above-average financial performance, a firm needs to consider not only the strategies which will build competitive advantage, but also the costs and extraneous conditions at the time of implementing those strategies [BAR 86b, DIE 89].

Although the RBV approach to strategy formulation is treated in the literature as generic in its applicability to all firms [MAR 03], Brush and Chaganti [BRU 98] note the lesser amount of empirical research on the strategic management of SMEs as compared to large firms. Notwithstanding this perceived research gap, there is however a general agreement in the literature that the distinctive organizational characteristics of SMEs mean they are likely to follow different paths from large firms in developing strategies for acquiring competitive advantage, and if SMEs are to compete successfully with large firms then they must acquire and develop capabilities based on those distinctive characteristics [ARA 08, TOR 12a, TOR 12b].

2.3. Proactive CSR in SMEs

Significantly, there is growing acknowledgment in the literature (see [BAN 05, TOR 12a, TOR 12b]) that in-line with the RBV approach value-creating strategies and competitive advantage in the future will be rooted in specific capabilities that deliver not only profitable outcomes, but which also actively support the economic, social and environmental principles of CSR. In line with the preceding discussion, we consider proactive CSR to be responsible business strategies adopted voluntarily by firms that simultaneously support sustainable economic, social and environmental development at a level above that required to comply with government regulation. As suggested by the European Commission [EUR 03], the way firms operate in the market is a crucial indicator of how they have integrated CSR and sustainability concerns into their mainstream organizational structure and decision-making processes. In line with Torugsa *et al.* [TOR 12b], we suggest the key to the successful integration of proactive CSR, and the unlocking of its potential for fortifying SME competitive advantage and financial performance lies in properly understanding the interactive nature of the relationships that operate between its three dimensions.

2.3.1. *Economic dimension of proactive CSR*

With the aim of supporting economic growth and prosperity, the economic dimension of proactive CSR is the means by which firms can pre-empt economic responsibility issues that might come up in their interactions with customers, suppliers and shareholders in the marketplace [EUR 03]. Economic-related proactive CSR goes beyond short-term profit maximizing and compliance issues to emphasize long-term economic performance, through the effective creation and distribution of goods and services that help increase the standard of living of the whole economy on a local, national and international level. It includes encouraging innovation, efficiency and wealth creation, which will in turn foster longevity in corporate profitability and growth [BAN 05, RUS 07, WIL 05].

The importance of economic-related proactive CSR for "value-creation" is well established in the literature with evidence being presented that firms can create and increase value by producing new and different products desired by consumers or by lowering the costs of inputs [BAN 05, POR 85, TOR 12a, TOR 12b]. Put simply, when the firm sells its products for a price that exceeds the cost of production, the firm can capture the extra value it creates and so enhance its financial performance. Not only can such value be captured by the firm, but it can also be distributed to consumers through products or services, to shareholders through dividends or equity and to employees through salaries [BAN 05]. Moreover, many researchers contend that SMEs, which they credit as sources of innovation, employment and wealth creation, are well positioned to benefit from economic-related CSR (see [LEP 06, WEN 99]). However, to achieve high value-related outcomes, effective management of economic capital and a long-term perspective in management and decision-making are needed; such an approach is needed in order to ensure that, at any time, cashflow sufficient to ensure liquidity while producing a persistent above-average return to shareholders can be achieved [DYL 02].

Perhaps, the most common form of SMEs is the owner-managed firm where ownership, decision-making and managerial control rest with the owner–manager [JEN 09]. In general, SME owner–managers are aware that their economic viability is crucially dependent on strong, long-term customer and supplier relationships built on loyalty, openness, honesty and fairness [HOR 94, LAH 05]. Where this awareness is acute and is coupled with a capacity to adapt flexibly and speedily in order to take advantage of changing market conditions and opportunities [GOF 95], SMEs are better equipped, when compared to larger firms, to benefit from economic-related proactive CSR. Empirical evidence also reveals that SMEs managed with a long-term perspective are those enterprises that achieve higher sales growth and profitability [BER 97, GIB 05]. Nevertheless, the adoption of a long-term perspective can prove difficult for some SME owner–managers, for whom management is primarily an adaptive process concerned with adjusting and manipulating a limited amount of resources to gain maximum immediate financial gain and ensure economic survival at least in the short term [BEA 00, SPE 99].

2.3.2. *Social dimension of proactive CSR*

The social dimension of proactive CSR has the workplace and the community as two points of focus in supporting social cohesion and equity. It actively recognizes "the health, safety and general well-being of employees;

motivate[s] the workforce by offering training and development opportunities; and enable[s] firms to act as good citizens in the local community" [EUR 03]. Through social-related proactive CSR, firms create a formal social dialog that enables the interests of all (internal and external) stakeholders to be considered with regard to CSR-related business issues; by so doing, mutually acceptable outcomes can result for the firm and its stakeholders [BAN 05].

Research examining the ability of SMEs to adopt social-related proactive CSR is inconclusive at this time. On the one hand, there is evidence that SMEs are influenced and affected by the general value systems which dominate social networks in their industry and in the value chain in which they operate; hence, norms and pressures from employees, peer firms and the community can drive SMEs to engage in socially responsible behaviors [ARB 97], and so enhance their reputations, images and competitiveness as employers and participants in the local community. By engaging in social-related proactive CSR, SMEs can develop close and positive relationships with their employees to address issues, such as workplace flexibility, democracy, diversity and equal opportunities, as well as with local communities through corporate citizenship initiatives. Such relationships, however, are not commonly expressed in formally written codes of conduct as in large firms; rather, they are characterized by high levels of informality, personal knowledge and familial ties that engender trust and reciprocity in interaction [LAH 05, WOR 06]. On the other hand, there is research highlighting the substantial difficulties SMEs experience in proactively developing social-related CSR. The restrictive impact of financial and human resource constraints on the offering of employee health and safety programs, training and development opportunities for employees and community involvement through charitable donations or sponsorship is well documented in the literature [BRA 06, GER 00, LEP 06].

2.3.3. *Environmental dimension of proactive CSR*

With the aim of supporting environmental integrity and protection, the environmental dimension of proactive CSR goes beyond regulatory compliance to focus on: (1) eco-efficiency and pollution prevention with innovative processes, products and operations aimed at reducing waste and energy consumption; and (2) environmental leadership with products and processes, and even business models, designed to minimize the firm's ecological footprint along the entire product lifecycle [ARA 08, BUY 03, SHA 98]. Environment-related proactive CSR often involves the adoption of internationally compliant environmental management systems and/or a total quality management approach to ensure the environmental impacts from a firm's operational activities are audited, monitored and managed systematically rather than on an *ad hoc* basis. Such systematic environmental management helps build a firm's creditability with external stakeholders and assists with embedding the environmental principle of CSR internally [WAL 94].

However, given that it can sometimes require quite sophisticated management expertise and integration of value chain activities beyond the ken of owner–managers, many SMEs face more difficulty than large firms in pursuing environment-related proactive CSR [RUT 00, TIL 99, SCH 02]. For instance, the decision to recover used product packaging from a firm's customers often requires the introduction of new value chain management practices (e.g. logistics); materials recycling programs often require high integration of multiple value chain components (e.g. production, inbound and outbound logistics, and post-sales service); and development of new environmentally friendly products can require cooperation between research and development and logistics value chain activities in order to identify sustainable and more efficient production materials and processes. Indeed, the complex nature of the management task of integrating value chain activities to support environment-related proactive CSR reinforces the need for the commitment of properly skilled human resources [SHA 98]. In particular, employee attitudes and environmental values, along with their technical and managerial skills, are of key importance. Unfortunately, these resources and skills are simply not always available in many SMEs [TIL 99]. As stated earlier, SMEs generally focus on strategies for achieving economic viability and prefer short-term investments that yield a high return; environment-related proactive CSR that requires significant investment without any certainty of returns in the short term is thus likely to find little favor as a competitive strategy for SMEs when simply achieving compliance represents a resource challenge [HOO 95, PET 00, TIL 99].

It would be remiss on our part to conclude this brief discussion of SMEs and environment-related proactive CSR without mentioning that a small number of empirical studies [ARA 08, HIL 00] have produced findings that support the more positive contrary view that SMEs can successfully engage in environment-related proactive CSR. A pertinent finding in this research [BIA 98] has been that successful adoption of environmentally responsible strategies by SMEs is often accompanied by the establishment of positive and stable relationships with external stakeholders (e.g. public institutions, research centers, industrial unions and government agencies). SMEs have been shown to be able to access through such relationships, the high-level skills, resources and information necessary for the introduction and management of complex environmental initiatives.

2.3.4. *Interaction and its contribution to SME financial performance*

As suggested by Becker-Olsen *et al.* [BEC 06], a trend toward a more socially acceptable and environmentally friendly form of consumerism is driving a transition in business from a narrow focus on economic-related CSR toward a broader focus incorporating social- and environment-related CSR. The emergence of this trend reflects the willingness of concerned consumers to pay a premium for socially and environmentally friendly products, and to exert market pressure by boycotting the products of firms with a poor reputation for social and environmental responsibility [GIE 11, GRO 11]. It also raises the possibility that financial institutions may perceive those SMEs with a poor social and environmental record as higher risk investment propositions, and so may refuse to extend new loans thereby impairing long-term profitability. These pressures have stimulated research interest not only in the contribution of proactive CSR to firm performance, but also in the interaction between the economic, social and environmental dimensions of proactive CSR and its link to firm performance.

To date, research specifically focusing on the contribution of proactive CSR to firm financial performance has been done mostly in the large firm context, and findings have generally been inconclusive. While some studies have shown no relationship [GIL 00, THO 03] and some have found a negative relationship [WAG 02], the majority of studies have revealed that proactive CSR is associated with enhanced product differentiation, improved production efficiencies, lower operation costs and reduced risk exposures; each of which contributes positively to a firm's financial performance [HAR 99, HAR 96, MAC 07]. The generally inconclusive nature of the findings might be explained in part by the difficulty of demonstrating empirically a clear causal effect of proactive CSR on financial performance, particularly as: some social-related CSR is philanthropic in nature and not necessarily related directly to the firm's operational business; and some environment-related CSR requires a substantial investment of resources for which returns cannot be realized in the short term. In the SME context, as already mentioned, the debate over whether a lack of resources constrains the implementation of proactive CSR thus limiting any associated financial benefits remains of ongoing research interest [GAD 08, ORL 01]. However, the existence of a positive link between proactive CSR and financial performance in SMEs has some empirical support [HAM 09].

Turning now to the possible interaction between the three dimensions of proactive CSR, clearly, there is justification for expecting such an effect. For example, successfully minimizing a firm's ecological footprint depends on the involvement of employees in the full range of value chain activities directly controlled by firms [HAR 95, NEH 98, RAM 00]. Social-related CSR, such as respecting and valuing all employees through the provision of training and employee development opportunities, complements environment-related CSR by building employees' awareness of and commitment to environmental values, as well as by improving the necessary technical and managerial skills [GRA 03]. The linkage to economic-related CSR is shown by the way properly designed social and environmental programs can trigger innovations that lower product cost and enhance the value proposition offered to customers [POR 95].

Of the very limited available research, we have the most recently presented evidence that supports the existence of an interaction effect and its implications for firm financial performance [TOR 12b]. Reporting an empirical study of 171 Australian manufacturing SMEs, we demonstrate that proactive CSR can provide significant scope for SME competitive advantage and financial performance improvement if its economic, social and environmental dimensions are adopted in an integrated and synergistic manner. Specifically, we reported that each dimension of proactive CSR influences SMEs' financial performance differently and that only the economic dimension was shown to have a direct positive association with financial performance, while no direct association was observed for either social or environmental dimension. Most importantly, our study also shows that the association between each individual CSR dimension and financial performance is positively moderated by the interaction of all three CSR dimensions, thus demonstrating that adoption of both social- and environment-related CSR is necessary for SMEs to maximize the contribution of economic-related CSR to financial performance. In other words, simultaneously incorporating economic-, social- and environment-related CSR is essential if SMEs wish to remain financially competitive, and not only survive but also prosper over the long term.

2.4. Capabilities for proactive CSR in SMEs

While the conventional wisdom in the literature is that a lesser and limited resource base can impose barriers to SMEs making substantial investments in proactive CSR [BAN 05, BRA 06, PAL 00], we argue that such constraints need not prevent SMEs from successfully adopting proactive CSR. Empirical evidence from recent studies [ARA 08, TOR 12a, TOR 12b] shows the probability that in the SME context the adoption of proactive CSR can be related to specific strategic capabilities built on the distinctive organizational characteristics of SMEs.

According to Yu [YU 01], entrepreneurial alertness and simple structures are two such characteristics. With smaller numbers of employees, coordination issues are less of a problem, and the influence of the SME owner–manager's vision on the direction of the firm is often greater than in large firms. In addition, SMEs often have a more unified culture, a stronger collective identity and better social relationships, when compared to large firms. The explanation for these characteristics lies in part in the tendency of SME owners to employ staff who they perceive to hold similar values to their own, thus providing more straightforward synergies in internal communication and sharing of firm-specific routines [YU 01]. On this basis, human relationships can significantly facilitate organizational efficiency and flexibility in SMEs. When compared to large firms, the flexible management structures of SMEs provide more responsiveness to changes in the marketplace and provide potentially greater innovativeness with which to respond with agility to competitors' actions [CHE 95, DEA 98, RAN 99]. Given the often limited resource and knowledge base available to SMEs generally, and the high level of dependence this creates upon external knowledge sources, technological developments, flexibility, responsiveness and innovativeness therefore become important characteristics for SMEs as sources of competitive advantage [JON 03]. Integrating and exploiting these characteristics can provide an effective basis for proactive CSR in SMEs that can deliver improvement in financial performance and a strategic competitive advantage [TOR 12a, TOR 12b].

Drawing on RBV theory that views organizational capabilities as a key consideration in formulating strategy, in the following we identify six specific capabilities (three internally directed and three externally directed), which can make adoption of proactive CSR in SMEs more likely. We suggest that when these strategic capabilities are leveraged for proactive CSR by SMEs, a competitive advantage and a superior financial performance can result.

2.4.1. *Internally directed capabilities*

2.4.1.1. *Shared vision*

Our recent research [TOR 12a] suggests that those SMEs which successfully develop a "Shared Vision" capability are better able to accumulate and harness the resources and skills necessary for adopting proactive CSR across its economic, social and environmental dimensions. A shared vision capability exists when the members of a firm have a shared feeling that the firm's objectives are important and that all of its members may contribute to defining them. To be clear, such a capability is not created simply by informing employees of the owner–managers' goals and objectives; rather, it arises when shared responsibility is accepted by all members of the firm for formulating and achieving an overall set of goals and objectives for the firm [GRA 03]. When a firm's members have the same perceptions about the firm's objectives, communication is made more efficient, with fewer misunderstandings about purpose and more opportunities to exchange ideas and resources [TSA 98]. Given its inherent requirement of employee involvement, a shared vision capability is fundamental to the different parts of a firm being able to bond together and integrate or combine resources effectively, and generate the internal pressure, commitment and support for innovation and change necessary for proactive CSR.

The smaller scale of SMEs and the coincidence of ownership and management control that typifies SME structures provide a context within which individual and organizational relationships frequently merge and become mutually reinforcing [JEN 09, WOR 06]. Given that owner–managers can build upon their personal and social ties, the task of developing a shared vision capability is made more straightforward in SMEs than in larger complex structures. The simpler management structures and shorter lines of communication generally characteristic of SMEs also facilitate greater involvement by all employees thus allowing the values and culture underpinning proactive CSR to be shared and embedded more easily across the entire firm [WOR 06]. Acting as a bonding mechanism that facilitates fluid communication and resource integration across functions and levels in a firm [TSA 98], a shared vision capability can help an SME determine which products present social and environmental responsibility risks thus contributing to economic gains in terms of improved product quality and safety.

While these benefits may be derived from a shared vision capability, we are not suggesting that every individual SME by definition has such a capability. Some SME owner–managers may lack the inclination and/or the necessary managerial skills and resources to work effectively with employees to develop shared objectives and foster behaviors supportive of proactive CSR [MER 95]. Moreover, employees who perceive proactive CSR as either unnecessary, too complex, or an unwelcome addition to already full workloads may resist its adoption. We believe that only those SMEs that have the resources and requisite managerial skills to exploit a shared vision of sustainable business, and create a climate of commitment and involvement of employees, will tend to adopt proactive CSR [TOR 12a, TOR 12b].

2.4.1.2. *Organizational learning*

Proactive CSR requires effective and efficient internal networks and communications. It also requires employees who understand how societal responsibility fits into the firm's operations and activities, and who have the training, knowledge and skills necessary to carry out proactive CSR as part of their work. In this respect, it is unlikely that competitive advantage built on the proactive CSR principles can be sustained in the longer term without a commitment to learning, managing and disseminating new knowledge widely within the firm. Organizational learning defined in this way as a capability [ARG 96], therefore, plays a crucial role in making SMEs successful with proactive CSR.

Research suggests that decision-making heavily dependent on individual managers' judgment, internal routines and relationships, and organizational culture can influence learning capability in SMEs [LIA 03]. SME owner–managers are often both driver and implementer of CSR-related values, thus making the exploration and exploitation of learning resources more straightforward. Less hierarchical management structures in SMEs, usually with either one individual or a very small team being responsible for a broad range of functional activities, can result in more opportunities for direct communication and learning through shared experiences. Fluid and open communications and a strong organizational culture allow learning outcomes to be embedded across the organization and CSR information to be rapidly disseminated. On the other hand, although SMEs have the ability to communicate well, they often have more difficulty than large firms in fostering learning through provision of formal training [KIT 03]. However, some researchers [FOR 04, PAT 00] contend that informal training is more frequently the chosen mechanism for learning in SMEs, and that the difficulties of measuring informal training lead to it being underreported with organizational learning capability in SMEs thus being underestimated.

2.4.1.3. *Capital management*

Poor management of capital reduces a firm's liquidity and ability to invest in productive assets, such as machinery and equipment, because funds are tied up in idle assets. Capital management is a capability that enables a firm to sustain the appropriate balance between its debt and asset management strategies and ensure a firm's survival generally, and specifically to underwrite the successful adoption of proactive CSR. As pointed out by Bansal [BAN 05], firms possessing a capital management capability more frequently attempt to develop best practices to avoid expensive capital refits associated with changing social and environment regulations than those firms without such a capability.

While a scarcity of slack resources can make it difficult for SMEs to develop a capital management capability, we argue that such scarcity is not necessarily a prohibitive factor. While SMEs often use informal controls and have little or no formal accounting expertise to implement and monitor capital planning and reporting, this informality of governance and control can have positive aspects for SMEs. For instance, informal controls and simple management structures and information systems can encourage internal connectivity or intensive dialogue between managers and all staff within an SME. This in turn can lead to the identification of opportunities for innovation which, when managed effectively, can lead to financial performance and capital growth through new and improved products, processes and operations [COY 12]. In addition, the efficient use of readily available low-cost internet-based tools that enable effective management of a firm's financial portfolio, as well as provide increased control over cash positions, means the development of a capital management capability becomes possible for all SMEs. Those SMEs which build and exploit a capital management capability are more likely to link proactive CSR with the overall business strategy and thus gain and sustain a competitive advantage as a consequence.

2.4.2. *Externally directed capabilities*

2.4.2.1. *Stakeholder engagement*

Stakeholder engagement is a capability that enables a firm to "establish trust-based collaborative relationships with a wide variety of stakeholders, especially those with non-economic goals" [SHA 98]. Because the interconnectedness of a firm and all its stakeholders is an integral element in the concept of CSR, the successful implementation of proactive CSR requires a firm to have the capability for building bridges with its stakeholders (through formal or informal engagement practice) based on the pursuit of shared interests, and to convince them to support the firm's chosen strategic actions [WHE 02]. Recent empirical research [ARA 08, TOR 12a] suggests that, like large firms, SMEs that recognize a wide variety of stakeholders are more likely to adopt proactive CSR than those that focus on a narrow range of stakeholders. A broadly focused stakeholder engagement capability supports active collaboration with the various stakeholder types through which CSR concerns might be managed and reduced.

Given their more limited internal resources, the challenge for SMEs is how to develop the capability with which the different and sometimes conflicting goals, priorities and demands of all stakeholders can be managed effectively [WER 99]. Unlike large firms that have the resources to utilize media and publicity to inform and engage stakeholders, SMEs are more likely to derive a stakeholder engagement capability from the closer relationships they can establish with stakeholders as a consequence of their flexibility and responsiveness to change in the business environment and stakeholder needs [JEN 06, JON 03]. Indeed, these characteristics enable SMEs to prioritize and focus more closely on particularly important external stakeholder relationships. This is a point of significance, given that implementation of complex responsibility strategies often requires specialized information on regulatory and social trends not generally available within an SME (i.e. cultivating a relationship of trust with customers and suppliers enables SMEs to better understand product lifecycles, design sustainably friendly products and consequently benefit from greater product acceptance). Building an enduring and mutually beneficial relationship with stakeholders via trust-based engagement can arguably help expand an SME's resources for undertaking proactive CSR.

2.4.2.2. *Strategic proactivity*

Strategic proactivity is a capability that enables a firm to initiate long-term policies and actions that shape the general business environment to its own advantage, and identify and capitalize on new opportunities emerging from changes in its business environment, rather than merely responding in a reactive fashion to those changes [ARA 98, SHA 07]. The concept of strategic proactivity as a capability has its roots in the study of Miles and Snow [MIL 78] indicating that strategically managed firms (or "prospectors") develop three types of processes that exploit external information and opportunities: *entrepreneurial* (analyzing all aspects of business context and growing by developing new products and markets), *engineering* (having flexible technologies capable of responding quickly to change) and *administrative* (formulating plans to reduce uncertainty and designing internal innovative processes). Strategy scholars contend that firms with a strategic proactivity capability often design internal processes that empower individuals to engage actively in CSR-oriented innovation, which assists in progressing toward a competitive advantage more quickly [ARA 98, STA 95].

Our research along with that of others [ARA 08, SHA 07] suggests that being entrepreneurial and seeking niche strategies in competitive markets can aid SMEs in the development of a strategic proactivity capability and hence the successful implementation of proactive CSR [TOR 12a, TOR 12b]. Because SMEs are motivated to seek opportunities to survive and prosper, they have a more immediate need, when compared to large firms generally, to act aggressively in the market and to initiate competitive actions to challenge the *status quo* [CHE 95]. On this basis, the creativeness, innovativeness and risk-taking behaviors characteristic of SMEs enable forecasting of external opportunities and threats, and proactively taking advantage of new niche markets for products and services with added value in the form of economic, social and environmental benefits [JEN 06].

2.4.2.3. *Continuous innovation*

Sharma *et al.* [SHA 07] define continuous innovation as a capability embedded in a firm's processes and routines for ongoing improvement, reconfiguration and invention of products, processes and organizational models. Given the market pressures on firms to imitate competitors' successful innovations, the capability of continuous innovation is thus fundamental for SMEs to maintain a competitive advantage through proactive CSR. Innovation can be either incremental or radical in form. For an SME to be innovative does not just involve the adoption of novel technology and processes or the introduction of products that are new to the market. Focusing on a range of incremental innovations – that lead to a better way to compete, that encompass the entire workforce of the firm and that add value to the existing products and processes to reduce their negative social and environmental impacts – is possibly a more effective way of enhancing SMEs' overall competitiveness [SHA 07].

Flexible specialization (the production in limited quantities of specially designed products of a specific quality for niche markets) can enable SMEs to link production networks with continuous improvement and superior innovative performance [FIE 91, VAN 95]. Flexible specialization, close collaboration with customers, suppliers and other firms in the marketplace, and managerial environments unburdened by excessive bureaucratic constraints can enable SMEs to take effective advantage of and rapidly respond to market needs. For example, customers can provide rapid feedback on technological solutions, and suppliers can help formulate technological design standards, all creating an environment that can drive innovation forward. New information and communication technologies that facilitate global reach at a low cost can assist SMEs to overcome disadvantages brought about by small scale and geographic location. For these reasons, we suggest that resource limitations do not put a continuous innovation capability beyond the reach of many SMEs.

2.5. Conceptual model for successful engagement in proactive CSR in SMEs

Drawing on RBV theory [BAR 91, GRA 91], and the extant empirical research discussed above, we present a theoretical model of the relationship between SME characteristics, CSR-related capabilities, proactive CSR, competitive advantage and financial performance (see Figure 2.1). According to this model, capabilities that derive from the distinctive organizational characteristics of SMEs provide a foundation for the implementation of proactive CSR that in turn promotes financial performance and competitive advantage. The model also suggests that to achieve superior financial performance, SMEs need to: consider what specific CSR-related strategies will build competitive advantage, identify and address the resource gaps inhibiting capability development and consider extraneous conditions at the time of implementing proactive CSR. More specifically, the model suggests the following:

– Smaller size, a common proxy for the unavailability of slack resources, is a relevant but not deterministic factor for the adoption of proactive CSR in SMEs.

– The development and exploitation of strategic capabilities must be guided by an understanding of the trends and changes in society and environment, and be undertaken in a manner that facilitates ongoing development of idiosyncratic firm attributes.

– SMEs have a set of distinctive organizational characteristics that can allow them to transcend size-related resource limitations and develop specific strategic capabilities which support the implementation of proactive CSR.

– Engaging in proactive CSR by leveraging three internally directed capabilities – shared vision, organizational learning and capital management.

– In tandem with three externally directed capabilities – stakeholder engagement, strategic proactivity and continuous innovation – can lead to the achievement of a competitive advantage and superior financial performance by SMEs.

– Only the economic dimension of CSR associates "directly" with financial performance, and importantly that association is made more positive through the interaction between the economic, social and environmental dimensions of CSR in SMEs.

– The social and environmental dimensions of proactive CSR need to be adopted in conjunction with the economic dimension of proactive CSR if cost-based and/or differentiation-based competitive advantages are to be properly gained. Where resource constraints limit full adoption of the social and environmental dimensions of proactive CSR, the particular subset of social- and environment-related CSR for which an SME is best equipped should be adopted in support of economic-related CSR.

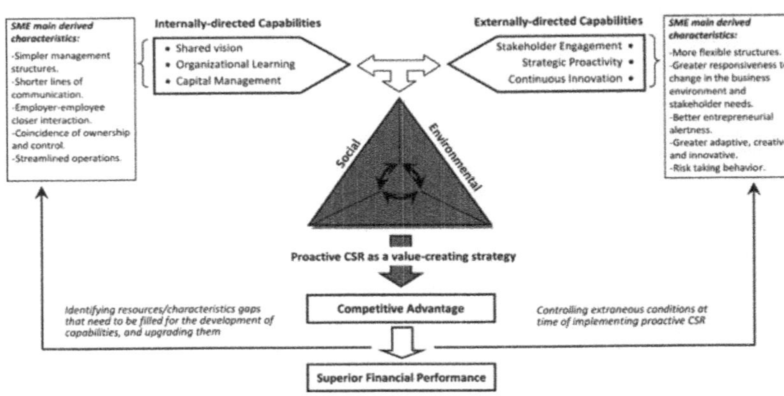

Figure 2.1. *A resource-based model of strategic capabilities for proactive CSR in SMEs*

2.6. Conclusion

This chapter invites researchers to reconsider the traditional assumption that the absence of slack financial and human resources prevents SMEs from successfully engaging in proactive CSR. Although resource constraints can often make it more difficult for SMEs to do so proactively, we argue that such constraints are not the determining factor that prevents SMEs from creating positive business outcomes generally, and specifically from improving financial returns through incorporating proactive CSR within the core of business strategy as part of "business as usual".

The model proposed in this chapter highlights the importance of specific strategic capabilities – that facilitate the integration of collective values about organizational goals (*shared vision*), enhance and disseminate new knowledge widely within the firm (*organizational learning*), manage and control short-term assets and liquidity (*capital management*), create sound trust-based relationships with stakeholders (*stakeholder engagement*), support the exploitation of new business opportunities (*strategic proactivity*) and continuously improve existing, or create new, innovations (*continuous innovation*) – as fundamental drivers for SME adoption of proactive CSR. The first three capabilities are internally directed and can be derived from the SME characteristics of simpler management and ownership structures that promote closer interaction and communication within the firm. The SME characteristics of greater flexibility and responsiveness, creativeness and innovation and better entrepreneurial alertness form the main basis from which the externally directed capabilities can be derived. We suggest that these distinctive organizational characteristics, when exploited in an integrated way with other available resources to develop capabilities for CSR, provide a means through which size-related constraints might be overcome in pursuit of competitive advantage and an improvement in SME financial performance. Given that the association with financial improvement has been shown as not consistent across the three dimensions of CSR [TOR 12b], paying primary attention to economic-related CSR and selectively focusing on elements of social- and environment-related CSR that drive and support the economic dimension are of key importance to sustainable long-term financial success for SMEs. In this way, a specific business case for proactive CSR in SMEs can be articulated, and proactive CSR can be seen as a value-creating strategy offering SMEs with competitive advantage and superior performance.

While the idea of proactive CSR as a value-creating strategy may not be new, its application to SMEs – particularly in relation to how to adopt proactive CSR that simultaneously supports economic growth and prosperity, social cohesion and equity, and environmental integrity and protection in a strategic way that contributes to SME financial success – is in its infancy. Our model and the discussion presented here suggest several possible opportunities for future research. One opportunity would be to use different methodologies and empirically test the model for further understanding the linkages between CSR-related strategic capabilities derived from SME distinctive characteristics, proactive CSR, competitive advantage and superior performance in SMEs. Furthermore, since many of the benefits of proactive CSR initiatives can be intangible and hard to measure, quantifying and translating such benefits into monetary value presents a research challenge that needs to be addressed if the business case for proactive CSR is to be made more persuasive for SME owner–managers. Future research comparing large firms and SMEs would also deepen understanding of the distinctive challenges SMEs face when it comes to the proactive implementation of responsible business strategies across the economic, social and environmental dimensions. Such an outcome would be of high value to individual SME owner–managers and government policy makers alike.

2.7. Bibliography

[AMI 93] AMIT R., SCHOEMAKER P.J.H., "Strategic assets and organizational rent", *Strategic Management Journal*, vol. 14, no. 1, pp. 33–46, 1993.

[ARA 98] ARAGON-CORREA J.A., "Strategic proactivity and firm approach to the natural environment", *Academy of Management Journal*, vol. 41, no. 5, pp. 556–567, 1998.

[ARA 08] ARAGON-CORREA J.A., HURTADO-TORRES N., SHARMA S., *et al.*, "Environmental strategy and performance in small firms: a resource-based perspective", *Journal of Environmental Management*, vol. 86, no. 1, pp. 88–103, 2008.

[ARB 97] ARBUTHNOT J.J., "Identifying ethical problems confronting small retail buyers during the merchandise buying process", *Journal of Business Ethics*, vol. 16, no. 7, pp. 745–755, 1997.

[ARG 96] ARGYRIS C., SCHÖN D., *Organisational Learning II*, Addison-Wesley, Wokingham, MA, 1996.

[AUS 01] AUSTRALIAN BUREAU OF STATISTICS, Small business in Australia, Catalogue No. 1321.0, Australian Government Printers, Canberra, 2001.

[AUS 09] AUSTRALIAN FIELDWORK SOLUTION, *SME Sustainability Index: Business Owners Survey on Sustainability and the Environment*, Smart Askers, Australia, 2009.

[BAN 05] BANSAL P., "Evolving sustainably: a longitudinal study of corporate sustainable development", *Strategic Management Journal*, vol. 26, no. 3, pp. 197–218, 2005.

[BAR 86a] BARNEY J.B., "Types of competition and the theory of strategy: toward an integrative framework", *Academy of Management Review*, vol. 11, no. 4, pp. 791–800, 1986.

[BAR 86b] BARNEY J.B., "Strategic factor markets: expectations, luck, and business strategy", *Management Science*, vol. 32, no. 10, pp. 1231–1241, 1986.

[BAR 91] BARNEY J.B., "Firm resources and sustained competitive advantage", *Journal of Management*, vol. 17, no. 1, pp. 99–120, 1991.

[BAR 97] BARNEY J.B., *Gaining and Sustaining Competitive Advantage*, Addison-Wesley Publishing Company, Reading, MA, 1997.

[BAR 01] BARNEY J.B., WRIGHT M., KETCHEN D.J., "The resource-based view of the firm: Ten years after 1991", *Journal of Management*, vol. 27, pp. 625–641, 2001.

[BAR 07] BARRETT R., MAYSON S., "Human resource management in growing small firms", *Journal of Small Business and Enterprise Development*, vol. 14, pp. 307–320, 2007.

[BEA 00] BEAVER G., JENNINGS P., "Editorial overview: small business, entrepreneurship and enterprise development", *Strategic Change*, vol. 9, no. 7, pp. 397–403, 2000.

[BEC 64] BECKER G.S., *Human Capital*, Columbia University Press, New York, 1964.

[BEC 06] BECKER-OLSEN K.L., CUDMORE B.A., HILL R.P., "The impact of perceived corporate social responsibility on consumer behaviour", *Journal of Business Research*, vol. 59, no. 1, pp. 46–53, 2006.

[BER 97] BERMAN J.A., GORDON D.D., SUSSMAN G., "A study to determine the benefits small business firms derive from sophisticated planning versus less sophisticated types of planning", *The Journal of Business and Economic Studies*, vol. 3, no. 3, pp. 1–11, 1997.

[BHA 04] BHATTACHARYA C.B., SEN S., "Doing better at doing good: when, why, and how consumers respond to corporate social initiatives", *California Management Review*, vol. 47, no. 1, pp. 9–24, 2004.

[BIA 98] BIANCHI R., NOCI G., "Greening SMEs competitiveness", *Small Business Economics*, vol. 11, no. 3, pp. 269–281, 1998.

[BRA 06] BRAMMER S., MILLINGTON A., "Firm size, organizational visibility and corporate philanthropy: an empirical analysis", *Business Ethics: A European Review*, vol. 15, no. 1, pp. 6–18, 2006.

[BRU 98] BRUSH C.G., CHAGANTI R., "Business without glamour? An analysis of resources on performance by size and age in small service and retail firms", *Journal of Business Venturing*, vol. 14, no. 3, pp. 233–257.

[BUY 03] BUYSSE K., VERBEKE A., "Proactive environmental strategies: a stakeholder management perspective", *Strategic Management Journal*, vol. 24, no. 5, pp. 453–470, 2003.

[CAR 79] CARROLL A.B., "A three-dimensional conceptual model of corporate performance", *Academy of Management Review*, vol. 4, no. 1, pp. 497–505, 1979.

[CHE 95] CHEN M., HAMBRICKD C., "Speed, stealth, and selective attach: how small firms differ from large firms in competitive behaviour", *Academy of Management Journal*, vol. 38, no. 2, pp. 453–482, 1995.

[COF 99] COFF R.W., "When competitive advantage doesn't lead to performance: the resource-based view and stakeholder bargaining power", *Organisation Science*, vol. 10, no. 2, pp.119–133, 1999.

[COL 98] COLLIS D.J., MONTGOMERYC A., *Corporate Strategy: A Resource-Based Approach*, Irwin/McGraw-Hill, New York, 1998.

[CON 91] CONNER K., "A historical comparison of resource-based theory and five schools of thought within industrial organization economics: do we have a new theory of the firm?", *Journal of Management*, vol. 17, no. 1, pp. 121–154, 1991.

[COY 12] COYTE R., RICCERI F., GUTHRIE J., "The management of knowledge resources in SMEs: an Australian case study", *Journal of Knowledge Management*, vol. 16, no. 5, pp. 789–807, 2012.

[DAY 94] DAY G.S., "The capabilities of market-driven organizations", *Journal of Marketing*, vol. 58, no. 4, pp. 37–52, 1994.

[DEA 98] DEAN T.J., BROWN R.L., BAMFORD C.E., "Differences in large and small firm responses to environmental context: strategic implications from a comparative analysis of business formations", *Strategic Management Journal*, vol. 19, no. 8, pp. 709–728, 1998.

[DIE 89] DIERICKXI I., COOL K., "Asset stock accumulation and sustainability of competitive advantage", *Management Science*, vol. 35, no. 12, pp. 1504–1511, 1989.

[DU 07] DU S., BHATTACHARYA C.B., SEN S., "Reaping relational rewards from corporate social responsibility: the role of competitive positioning", *International Journal of Research in Marketing*, vol. 24, no. 3, pp. 224–241, 2007.

[DUN 03] DUNPHY D., "Corporate sustainability: challenge to managerial orthodoxies", *Journal of the Australian and New Zealand Academy of Management*, vol. 9, no. 1, pp. 2–11, 2003.

[DYL 02] DYLLICK T., HOCKERTS K., "Beyond the business case for corporate sustainability", *Business Strategy and the Environment*, vol. 11, no. 2, pp. 130–141, 2002.

[EUR 03] EUROPEAN COMMISSION, Responsible Entrepreneurship: A Collection of Good Practice Cases among Small and Medium Sized Enterprises across Europe, Brussels, Belgium, 2003.

[FIE 91] FIEGENBAUM A., KARNANI A., "Output flexibility: a competitive advantage for small firms", *Strategic Management Journal*, vol. 12, no. 2, pp. 101–114, 1991.

[FIT 11] FITJAR R.D., "Little big firms? Corporate social responsibility in small businesses that do not compete against big ones", *Business Ethics: A European Review*, vol. 20, no. 1, pp. 30–44, 2011.

[FOR 04] FORTH J., BEWLEY H., BRYSON A., Small and Medium-Sized Enterprises: Findings from the 2004 Workplace Employment Relations Survey, Department of Trade and Industry, London, 2004.

[GAD 08] GADENNE D.L., KENNEDY J., MCKEIVER C., "An empirical study of environmental awareness and practices in SMEs", *Journal of Business Ethics*, vol. 10, pp. 1–19, 2008.

[GAL 72] GALE B.T., "Market share and rate of return", *The Review of Economics and Statistics*, vol. 54, no. 4, pp. 412–423, 1972.

[GER 00] GERRANS P., HUTCHINSON B., "Sustainable development and small to medium-sized enterprises: a long way to go", in HILLARY R. (ed.), *Small and Medium-Sized Enterprises and the Environment*, Greenleaf, Sheffield, pp. 75–81, 2000.

[GIB 85] GIBB A.A., SCOTT M., "Strategic awareness, personal commitment and the process of planning in the small business", *Journal of Management Studies*, vol. 22, no. 6, pp. 597–631, 1985.

[GIB 05] GIBSON B., CASSER G., "Longitudinal analysis of relationships between planning and performance in small firms", *Small Business Economics*, vol. 25, no. 3, pp. 207–222, 2005.

[GIE 11] GIELISSEN R.B., "Why do customers buy socially responsible products?", *International Journal of Business and Social Science*, vol. 2, no. 3, pp. 21–35, 2011.

[GIL 00] GILLEY K., WORELL D., EI-JELLY A., "Corporate environmental initiatives and anticipated firm performance", *Journal of Management*, vol. 26, pp. 1199–1216, 2000.

[GOF 95] GOFFEE R., SCASE R., *Corporate Realities: The Dynamics of Large and Small Organisations*, International Thomson Business Press, London, 1995.

[GRA 03] GRAAFLAND J., VAN DE VEN B., STOFFELE N., "Strategies and instruments for organising CSR by small and large businesses in the Netherlands", *Journal of Business Ethics*, vol. 47, no. 1, pp. 45–60, 2003.

[GRA 91] GRANT R.M., "The resource-based theory of competitive advantage: implications for strategy formulation", *California Management Review*, vol. 33, no. 3, pp. 114–135, 1991.

[GRA 98] GRANT R.M., *Contemporary Strategy Analysis: Concepts, Techniques, Applications*, Basil Blackwell, Cambridge, MA, 1998.

[GRE 04] GREENFIELD W.M., "In the name of corporate social responsibility", *Business Horizons*, vol. 47, pp. 19–28, 2004.

[GRO 11] GROZA M.D., PRONSCHINSKE M.R., WALKER, M., "Perceived organizational motives and consumer responses to proactive and reactive CSR", *Journal of Business Ethics*, vol. 102, pp. 639–652, 2011.

[HAL 93] HALL R., "A framework linking intangible resources and capabilities to sustainable competitive advantage", *Strategic Management Journal*, vol. 14, no. 8, pp. 607–618, 1993.

[HAM 09] HAMMANN E., HABISCH A., PECHLANER H., "Values that create value: socially responsible business practices in SMEs – empirical evidence from German companies", *Business Ethics: A European Review*, vol. 18, no. 1, pp. 37–51, 2009.

[HAR 99] HARRISON J.G., FREEMAN R.E., "Stakeholders, social responsibility and performance: empirical evidence and theoretical perspectives", *Academy of Management Journal*, vol. 42, no. 5, pp. 479–485, 1999.

[HAR 95] HART S.L., "A natural-resource-based view of the firm", *Academy of Management Review*, vol. 20, no. 4, pp. 986–1014, 1995.

[HAR 96] HART S.L., AHUJA G., "Does it pay be green? An empirical examination of the relationship between emission reduction and firm performance", *Business Strategy and the Environment*, vol. 5, pp. 30–37, 1996.

[HAR 04] HART S.L., SHARMA S., "Engaging fringe stakeholders for competitive imagination", *Academy of Management Executive*, vol. 18, no. 1, pp. 7–18, 2004.

[HAW 06] HAWKINS D.E., *Corporate Social Responsibility: Balancing Tomorrow's Sustainability and Today's Profitability*, Palgrave Macmillan, New York, 2006.

[HIL 00] HILLARY R., *Small and Medium-Sized Enterprises and the Environment*, Greenleaf Publishing, Sheffield, 2000.

[HOO 95] HOOPER P.D., GIBBSD C., Profiting from environmental protection: a Manchester business survey, Report for the Cooperative Bank, Manchester, 1995.

[HOR 94] HORNSBY J.S., KURATKO D.F., NAFFZIGGER, D.W., *et al.*, "The ethical perceptions of small business owners: a factor analytic study", *Journal of Small Business Management*, vol. 32, no. 4, pp. 9–17, 1994.

[JEN 06] JENKINS H., "Small business champions for corporate social responsibility", *Journal of Business Ethics*, vol. 67, no. 3, pp. 241–256, 2006.

[JEN 09] JENKINS H., "A business opportunity model of corporate social responsibility for small and medium-sized enterprises", *Business Ethics: A European Review*, vol. 18, no. 1, pp. 21–36, 2009.

[JON 03] JONES O., "Competitive advantage in SMEs: towards a conceptual framework", in JONES O., TILLEY F. (eds), *Competitive Advantage in SMEs: Organising for Innovation and Change*, Wiley, Chichester, pp. 15–33, 2003.

[KIT 03] KITSON M., WILKINSON F., "Labour mobility, training and labour force flexibility", in COSH A., HUGHES A. (eds), *Enterprise Challenged: Policy and Performance in the British SME Sector 1999–2002*, Centre for Business Research, Cambridge, pp. 32–44, 2003.

[KLA 99] KLASSEN R.D., WHYBARK D.C., "The impact of environmental technologies on manufacturing performance", *Academy of Management Journal*, vol. 42, no. 6, pp. 599–615, 1999.

[KOT 05] KOTLER P., LEE N., *Corporate Social Responsibility: Doing the Most Good for Your Cause*, John Wiley, Hoboken, NJ, 2005.

[LAH 05] LAHDESMAKI M., "When ethics matters – interpreting the ethical discourse of small nature-based entrepreneurs", *Journal of Business Ethics*, vol. 61, no. 1, pp. 55–68, 2005.

[LEP 06] LEPOUTRE J., HEENE A., "Investigating the impact of firm size on small business social responsibility: a critical review", *Journal of Business Ethics*, vol. 67, no. 3, pp. 257–273, 2006.

[LIA 03] LIAO J., WELSCH H., STOICA M., "Organizational absorptive capability and responsiveness: an empirical investigation of growth-oriented SMEs", *Entrepreneurship Theory and Practice*, vol. 28, pp. 63–85, 2003.

[LIN 10] LINDGREEN A., SWAEN V., "Corporate social responsibility", *International Journal of Management Reviews*, vol. 12, no. 1, pp. 1–7, 2010.

[MAC 07] MACKEY A., MACKEY T.B., BARNEY J.B., "Corporate social responsibility and firm performance: investor preferences and corporate strategies", *Academy of Management Review*, vol. 32, no. 3, pp. 817–835, 2007.

[MAN 02] MAN T.W.Y., LAU T., CHAN K.F., "The competitiveness of small and medium enterprises: a conceptualization with focus on entrepreneurial competencies", *Journal of Business Venturing*, vol. 17, no. 2, pp. 123–142, 2002.

[MAR 03] MARSDEN A., FORBES C., "Strategic management for small and medium-sized enterprises (SMEs)", in JONES O., TILLEY F. (eds), *Competitive Advantage in SMEs: Organising for Innovation and Change*, Wiley, England, pp. 34–53, 2003.

[MAR 98] MARSHALL L., *Economic Instruments and the Business Use of Energy*, Stationery Office, London, 1998.

[MAT 08] MATTEN D., MOON J., "Implicit and explicit CSR: a conceptual framework for a comparative understanding of corporate social responsibility", *Academy of Management Review*, vol. 33, no. 2, pp. 404–424, 2008.

[MCW 06] MCWILLIAMS A., SIEGEL D.S., WRIGHT P.M., "Corporate social responsibility: strategic implications", *Journal of Management Studies*, vol. 43, pp. 1–18, 2006.

[MER 95] MERZ G.R., SAUBER M., "Profiles of managerial activities in small firms", *Strategic Management Journal*, vol. 16, no. 7, pp. 551–564, 1995.

[MIL 78] MILES R., SNOW C., *Organizational Strategy, Structure and Process*, McGraw Hill, New York, 1978.

[MOO 99] MOORMAN C., SLOTEGRAAF R.J., "The contingency value of complementary capabilities in product development", *Journal of Marketing Research*, vol. 36, pp. 239–257, 1999.

[NEH 98] NEHRT C., "Maintainability of first mover advantages when environmental regulation differs between countries", *Academy of Management Review*, vol. 23, no. 1, pp. 77–97, 1998.

[ORL 01] ORLITZKY M., "Does firm size confound the relationship between social performance and firm financial performance?", *Journal of Business Ethics*, vol. 33, no. 2, pp. 167–180, 2001.

[PAL 00] PALMER J., "Helping small and medium-sized enterprises improve environmental management: Lessons from proactive small and micro firms", in HILLARY R. (ed.), *Small and Medium Sized Enterprises and the Environment*, Greenleaf Publishing, Sheffield, pp. 325–342, 2000.

[PAT 00] PATTON D., MARLOW S., HANNON P., "The relationship between training and small firm performance: research frameworks and lost quests", *International Small Business Journal*, vol. 19, no. 1, pp. 11–27, 2000.

[PEN 59] PENROSE E.T., *The Theory of the Growth of the Firm*, Wiley, New York, 1959.

[PET 93] PETERAF M.A., "The cornerstones of competitive advantage: a resource-based view", *Strategic Management Journal*, vol. 14, no. 3, pp. 179–191, 1993.

[PET 03] PETERAF M.A., BARNEY J.B., "Unraveling the resource-based tangle", *Managerial and Decision Economics*, vol. 24, no. 4, pp. 309–323, 2003.

[PET 00] PETTS J., "Smaller enterprises and the environment: organisational learning potential", in FINEMAN S. (ed.), *The Business of Greening*, Routledge, London, pp. 153–168, 2000.

[POR 85] PORTER M.E., *Competitive Advantage: Creating and Sustaining Superior Performance*, Free Press, New York, 1985.

[POR 95] PORTER M.E., VAN DER LINDE C., "Green and competitive: Ending the stalemate", *Harvard Business Review*, vol. 73, no. 5, pp. 120–134, 1995.

[POW 01] POWELL T.C., "Competitive advantage: logical and philosophical considerations", *Strategic Management Journal*, vol. 22, no. 9, pp. 875–888, 2001.

[PRA 90] PRAHALAD C.K., HAMEL G., "The core competence of the corporation", *Harvard Business Review*, vol. 68, no. 3, pp. 79–91, 1990.

[RAM 00] RAMUS C.A., STEGER U., "The roles of supervisory support behaviors and environmental policy in employee 'ecoinitiatives' at leading-edge European companies", *Academy of Management Journal*, vol. 43, no. 4, pp. 605–626, 2000.

[RAN 99] RANGONE A., "A resource-based approach to strategy analysis in small-medium sized enterprises", *Small Business Economics*, vol. 12, no. 3, pp. 233–248, 1999.

[ROB 84] ROBINSON R.B., PEARCE J.A., "Research thrusts in small firm strategic planning", *Academy of Management*, vol. 9, no. 1, pp. 128–137, 1984.

[RUM 84] RUMELT R.P., "Towards a strategic theory of the firm", in LAMB B. (ed.), *Competitive Strategic Management*, Prentice Hall, Englewood Cliffs, NJ, pp. 556–570, 1984.

[RUM 91] RUMELT R.P., "How much does industry matter?", *Strategic Management Journal*, vol. 12, no. 3, pp. 167–185, 1991.

[RUM 94] RUMELT R.P., SCHENDEL D., TEECE D., *Fundamental Issues in Strategy*, Harvard University Press, Cambridge, MA, 1994.

[RUS 07] RUSSELL S.V., HAIGH N.L., GRIFFITHS A., "Understanding corporate sustainability: recognizing the impact of different governance systems", in BENN S., DUNPHY D. (eds), *Corporate Governance and Sustainability: Challenges for Theory and Practice*, Routledge, New York, pp. 36–56, 2007.

[RUS 97] RUSSO M.V., FOUTS P.A., "A resource-based perspective on corporate environmental performance and profitability", *Academy of Management Journal*, vol. 40, no. 3, pp. 534–559, 1997.

[RUS 09] RUSSO A., TENCATI A., "Formal vs. informal CSR strategies: evidence from Italian micro, small, medium-sized and large firms", *Journal of Business Ethics*, vol. 85, no. 2, pp. 339–353, 2009.

[RUT 00] RUTHERFOORD R., BLACKBURN R.A., SPENCE L.J., "Environmental management and the small firm: an international comparison", *International Journal of Entrepreneurial Behaviour and Research*, vol. 6, no. 6, pp. 310–325, 2000.

[SCH 02] SCHAPER M., "Small firms and environmental management: predictors of green purchasing in western Australian pharmacies", *International Small Business Journal*, vol. 20, pp. 235–251, 2002.

[SEX 85] SEXTON D.L., VAN AUKEN P., "A longitudinal study of small business strategic planning", *Journal of Small Business Management*, vol. 23, pp. 7–15, 1985.

[SHA 98] SHARMA S., VREDENBURG H., "Proactive corporate environmental strategy and the development of competitively valuable organizational capabilities", *Strategic Management Journal*, vol. 19, no. 8, pp. 729–753, 1998.

[SHA 07] SHARMA S., ARAGON-CORREA J.A., RUEDA-MANZANARES A., "The contingent influence of organizational capabilities on proactive environmental strategy in the service sector: an analysis of North American and European Ski Resort", *Canadian Journal of Administrative Science*, vol. 24, no. 4, pp. 268–283, 2007.

[SHU 85] SHUMAN J.C., SHAW J.J., SUSSMAN G., "Strategic planning in smaller rapid growth companies", *Long Range Planning*, vol. 18, no. 6, pp. 48–53, 1985.

[SMA 00] SMALLBONE D., WYER P., "Growth and development in the small firm", in CARTER S., JONES-EVANS D. (eds), *Enterprise and Small Business: Principles, Practices and Policies*, Prentice Hall, Harlow, pp. 409–433, 2000.

[SMI 98] SMITH M.A., KEMP R., *Small Firms and the Environment: A Grounded Report*, Groundwork, Birmingham, 1998.

[SPE 99] SPENCE L.J., "Does size matter? The state of the art in small business ethics", *Business Ethics: A European Review*, vol. 8, no. 3, pp. 163–174, 1999.

[STA 76] STANWORTH M.J.K., CURRAN J., "Growth and the small firm- an alternative view", *Journal of Management Studies*, vol. 13, no. 2, pp. 95–110, 1976.

[STA 95] STARIK M., RANDS G.P., "Weaving an integrated web: multilevel and multisystem perspectives of ecologically sustainable organizations", *Academy of Management Review*, vol. 20, no. 4, pp. 908–935, 1995.

[STR 77] STRATEGIC PLANNING INSTITUTE, *Selected Findings from the PIMS Program*, Strategic Planning Institute, Cambridge, MA, 1977.

[THO 03] THORNTON D., KAGAN R.A., GUNNINGHAM M., "Sources of corporate environmental performance", *California Management Review*, vol. 46, no. 1, pp. 127–141, 2003.

[TIL 99] TILLEY F., "The gap between environmental attitudes and the environmental behaviour of small firms", *Business Strategy and the Environment*, vol. 8, no. 4, pp. 238–248, 1999.

[TIL 00] TILLEY F. "Small firm environmental ethics: how deep do they go?", *Business Ethics: A European Review*, vol. 9, no. 1, pp. 31–41, 2000.

[TOR 12a] TORUGSA N., O'DONOHUE W., HECKER R., "Capabilities, proactive CSR and financial performance in SMEs: empirical evidence from an Australian manufacturing industry sector", *Journal of Business Ethics*, vol. 109, no. 4, pp. 483–500, 2012.

[TOR 12b] TORUGSA N., O'DONOHUE W., HECKER R., "Proactive CSR: an empirical analysis of the role of its economic, social and environmental dimensions on the association between capabilities and performance", *Journal of Business Ethics*, 2012.

[TSA 98] TSAI W., GHOSHAL S., "Social capital and value creation: the role of intrafirm networks", *Academy of Management Journal*, vol. 41, no. 4, pp. 464–476, 1998.

[UNI 02] UNITED NATIONS INDUSTRIAL DEVELOPMENT ORGANIZATION, Corporate social responsibility: implications for small and medium enterprise in developing countries, UNIDO Report, Vienna International Centre, 2002.

[VAN 95] VAN DIJK M.P., "Flexible specialisation, the new competition and industrial districts", *Small Business Economics*, vol. 7, no. 1, pp. 15–28, 1995.

[VOS 98] VOSSEN R.W., "Relative strengths and weaknesses of small firms in innovation", *International Small Business Journal*, vol. 16, no. 3, pp. 88–94, 1998.

[WAG 02] WAGNER M., VAN PHU N., AZOMAHOU T., WHERMEYER W., "The relationship between the environmental and economic performance of firms: an empirical analysis of the European paper industry", *Corporate Social Responsibility and Environmental Management*, vol. 9, pp. 133–146, 2002.

[WAL 94] WALLEY N., WHITEHEAD B., "It's not easy being green", *Harvard Business Review*, vol. 72, no. 3, pp. 46–52, 1994.

[WAR 85] WARTICK S.L., COCHRAN P.L., "The evaluation of the corporate social performance model", *Academy of Management Review*, vol. 10, no. 4, pp. 758–769, 1985.

[WEN 99] WENNEKERS S., THURIK R., "Linking entrepreneurship and economic growth", *Small Business Economics*, vol. 13, no. 1, pp. 27–55, 1999.

[WER 99] WERHANE P.H., FREEMAN R.E., "Business ethics: the state of the art", *International Journal of Management Reviews*, vol. 1, no. 1, pp. 1–16, 1999.

[WER 84] WERNERFELT B., "A resource-based view of the firm", *Strategic Management Journal*, vol. 5, no. 2, pp. 171–181, 1984.

[WHE 02] WHETTEN D.A., RANDS G., GODFREY P., "What are the responsibilities of business to society?" in PETTIGREW A., THOMAS H., WHITTINGTON R. (eds), *Handbook of Strategy and Management*, Sage, London, pp. 373–408, 2002.

[WIL 05] WILLARD B., *The Sustainability Advantage: Seven Business Case Benefits of a Triple Bottom Line*, New Society Publisher, Gabriola Island, 2005.

[WIL 75a] WILLIAMSON O.E., *Markets and Hierarchies: Analysis and Antitrust Implications*, Free Press, New York, 1975.

[WIL 75b] WILSON L., "What one company is doing about today's demands on business", in STEINER G.A. (ed.), *Changing Business– Society Interrelationships*, UCLA, Los Angeles, pp. 121–140, 1975.

[WOR 06] WORTHINGTON I., RAM M., JONES T., "Exploring corporate social responsibility in the UK. Asian small business community", *Journal of Business Ethics*, vol. 67, no. 2, pp. 201–217, 2006.

[YU 01] YU T.F., "Towards a capabilities perspective of small firm", *International Journal of Management Reviews*, vol. 3, no. 3, pp. 185–197, 2001.

Chapter 3

Innovative Management Development in the Automotive Supply Industry – A Preliminary Case Study for the Development of an Innovative Approach to Innovation Management

3.1. Introduction

In a globalized world, enterprises are confronted with an increasingly fast change of technical and technological possibilities, available knowledge and differentiated markets, customer demands, lifestyles and wishes of other economic agents. Triggered and accompanied by new options of information and communication technology, worldwide competition has increased. Against this background, organizations need to innovate in response to this changing environment in order to capitalize on opportunities offered by these changes in technology, marketplaces, structures and dynamics. The speed of economic change, however, can be seen as a chance for new business ideas, companies, markets or market segments [BAR 09, VOL 10a, VOL 10b]. The companies that can adapt faster to new situations than their competitors will be successful because they can create a temporary competitive edge. By doing that, they can temporarily suspend the neoclassical logic of the profit,

---

Chapter written by Frank E.P. DIEVERNICH and Kim Oliver TOKARSKI.

which will disappear in a state of equilibrium by continuously creating economic imbalances. Of course, such profitable imbalances have to be created over and over again in times of fast economic change and strong competition [DEV 98; VOL 10b]. In this context, the concept of innovation is perceived as a central component in creating value and sustaining competitive advantage [BAR 09]. Small and medium enterprises (SMEs) are generally considered to be innovative companies [VOL 10a]. We can find hidden champions among SMEs as well as among family companies. Entrepreneurship represents their very essence. Indeed, they were found to be innovative, if we believe the myths about invention originating in small garages and workshops [SIM 09].

The case study presented and discussed in this chapter does actually deal with a company founded in a garage over 30 years ago. Innovation is central to the philosophy of this SME and dictates what it thinks and does on a daily basis. For this company, innovation is normality. If, of course, those companies whose products are characterized by a focus on core technology wish to maintain and further develop this technology, it is essential that they anchor innovation as a cornerstone of their daily activity. Indeed, in this company, it appears as though innovation and operative business go hand in hand, whereby innovative activity solely relates to products and technologies. Innovation does not seem to have a separate organizational significance beyond the operational business [THO 06].

But what if this identity, this self-image, is no longer sufficient to meet the (innovative) requirements of the market? What if reliance on technological innovation and R&D departments is no longer enough to secure success in the future? What if innovation degenerates into a litany that is no longer capable of directing daily action within the company? What if innovation itself has become a blind spot, because there is a belief that innovation is rooted in the culture and will thus occur by itself? What if innovation and the innovation process is no longer seen as something innovative, but is merely following a well-worn path? Finally, what if a perception has developed within the company (this is usually an insidious process) that investment in innovation in the widest sense is a "fair weather" topic, and hence one which should only be addressed when the economic climate is benign? Our aim is to show how innovative management development, which should be part of innovative innovation management, can prevent the situation lapsing into the negative scenarios referred to above.

3.2. Innovation

Innovation can be seen as a complex construct [DAM 96]. It is regarded as a major source for a competitive advantage of an organization. But there is not a single success factor of innovation and innovation management. Interdependencies of innovation and other factors are complex. So, success of innovation is attributed to numerous factors, such as R&D expenditure, technological advancement, the organization's market orientation and capabilities. Another organizational success factor of innovation is how organizations or individuals view innovation or how organizations define innovation [GOS 05]. Therefore, it is important to shed some light on what innovation is.

When defining the concept of innovation, Kwaku Atuahene-Gima notes: "Get the definition wrong, and you've got the concept wrong" [ATU 12, p. 66]. But it is not easy to give a clear definition of the concept of innovation because no single definition of the term exists. Furthermore, the term "innovation" is often used rather heterogeneously [GRA 08]. The definition of the concept of innovation is of interest in theory (science) as well as in practice (industry and organizations). In an organizational context, it is believed that the way innovation is defined will constitute the business activities of an enterprise. Against this background, theory and practice often use a multiperspective approach to define innovation. Therefore, various definitions of innovation exist in literature and in organizational practice [GOS 05].

At first, the concept "innovation" has to be differentiated from the term "invention". An invention can be described as the discovery of a novel (technical) realization of a problem solution. So, inventions are a preliminary step to a marketable innovation. In this context, inventions can occur in a planned process (R&D) or unplanned process (accident) [VOL 10a]. There are numerous key inventions that were produced by accident and therefore unplanned (e.g. penicillin, cardiac pacemaker, saccharin and Teflon).

Inventions (ideas as well as prototypes), however, are not necessarily innovations. Within the innovation process, an invention has to be transformed into an innovation [VOL10a, VOL 10b, SCH 34, AES 09]. Against this background, the concept innovation can be defined as "a novel problem solution of a technical, economic, organisational or social type, which is realized in enterprises or in the market" [VOL 10a, p. 100]. So, the transformation of an invention into a (marketable) product can be regarded as an innovation.

Joseph A. Schumpeter already had a comprehensive understanding of the innovation concept at the beginning/middle of the 20th Century. Schumpeter put entrepreneurship (and the entrepreneur) at the same level as innovation. Therefore, innovation, entrepreneurship and the entrepreneur are highly interconnected. An entrepreneur is an innovator who implements a (entrepreneurial) change.

Schumpeter describes the concept of innovation (fostered through the entrepreneur) as the implementation and realization of continuous new factor combinations in the form of new products, new qualities of already existing successful products, new production methods, the development of new markets, new organizational structures or new procurement methods. According to Schumpeter, innovation is "the creative destruction of the existing by an entrepreneur" [SCH 34, SCH 42, VOL 10a]. The process of creative destruction can be defined as follows: "The opening up of new markets, foreign or domestic, and the organisational development [...] illustrate the same process of industrial mutation [...] that incessantly revolutionizes the economic structure from within, incessantly destroying the old one, incessantly creating a new one. This process of Creative Destruction is the essential fact about capitalism. It is what capitalism consists of and what every capitalist concern has got to live with" [SCH 42, p. 83]. This aspect of destruction is very important in and for market economies, because it helps to realize an optimum resource allocation and competition [VOL 10a, SCH 34]. So innovation, entrepreneurship and the entrepreneur are triggers for a creative destruction of products, markets and industries that foster change. On the basis of Schumpeter's thoughts, other selected definitions are presented to show the (broad) scope of a (modern) understanding of innovation.

According to the OECD Oslo manual, an "[...] innovation is the implementation of a new or significantly improved product (good or service), or process, a new marketing method, or a new organisational, method in business practices, workplace organisation or external relations" [OEC 05, p. 46]. This broad definition covers main aspects presented by Schumpeter. It seems to be important to note that not only external (market related) innovations are of interest but also internal (organizational) innovations.

The term "innovation" can also comprise product innovations, process innovations, market innovations, organizational innovations as well as social or cultural innovations [STE 07]. The field of innovation can furthermore be differentiated into various aspects such as (1) new products or technologies versus organizational innovation, (2) invention versus diffusion or adaptation, (3) meeting market demands versus capitalization of genuine talents and

(4) individual versus social innovations [HEI 98]. Furthermore, organizations innovate in business models and channels [CAR 99]. Against this background, Arreger defines innovation with regard to the environment of its emergence/origin as a significant change of the status quo of a social system (e.g. an organization) that is based on new findings, social behavior, materials or machines and aims at a direct and/or indirect improvement within or outside the system [ARR 76]. All in all, an innovation can be observed on different levels, for example on the level of the industry, organization or individual [DAM 96]. Therefore, a great number of different definitions exist that suit theory and practice.

However, a single innovation alone is not necessarily the main long-term success factor for an organization [TUS 97]. The long-term success of an organization may rely more on an overall innovation orientation at the organizational level. So, an innovation orientation generates capabilities that continuously foster innovations [SIG 06]. The concept of innovation is interconnected to change. So, innovation is used as a tool in organizations to influence their environment or react to changing environments [DAM 91]. But just focusing on innovation (in terms of invention) as a remedy for (sustainable) organizational development can be seen critically. Thus, [SAF 07] calls it the "cult of innovation". It is crucial that an organization has the ability to judge whether it is useful to invest resources to bring an invention to the market or not. In this context, it is important to resonate with (potential) customers and address their (unmet) needs and desires and not just the company's feeling that the organization has to innovate [SAF 07]. Companies therefore collaborate with customers for two main reasons. On the one hand, they search for new ideas. On the other hand, they search for new ways to reduce uncertainty that is associated with the introduction of innovations to the market [VON 88]. However, the management of the organization has to make the final decision in terms of how to proceed. Therefore, the management (entrepreneur) has to be aware of and has to have capabilities to perceive opportunities and innovation.

This is important because there is a positive correlation of how innovative an organization is and the educational level of its managers. So, an unskilled SME management reduces its innovative capability and vice versa [SUB 11]. Based on the above brief discussion, the following case study focuses on different organizational aspects to foster organizational innovation (capabilities). However, as noted before, it is very important to stress that there has to be a commitment from the entrepreneur or (top) manager/management team to foster an organizational orientation on innovation.

3.3. Case study

The following practical example should show how a medium-sized family business dared to organize itself in such a way that it could maintain innovative drive while anticipating and weathering crises, and thereby set itself up for success into the future. The company in question is a supplier to the automotive industry. It specializes in the production of cast resins and machine systems and is located in southern Germany. The majority of its customers are in the automotive (supplier) industry. The company is organized along functional lines. The group of companies has approximately 600 staff worldwide. The HR function had an important role to play. Management within this function had to go beyond just ensuring innovation could take root. It had to assume leadership of the innovative philosophy across the entire company, because engineering and R&D were themselves no longer the fount of innovation. This reality stood in stark contrast to the company's self-perception and verbal statements from those active in the operative business[1]. It was in fact the (family-run) management that gave HR this modern observational role to foster innovation. What makes this case study different from other companies and personnel functions was the fact that here a proactive decision had been made to focus on a link between HR and innovation, and this in turn lent a very modern aspect to the role HR had to play. In the broadest sense, this example shows that HR extended its focus from managing and "caring" for employees (personnel administration, personnel development, skills enhancement, career planning, etc.) to include the structural development of the overall organization. This added a new detail to the HR remit – it had to assume an "innovation-oriented correction function" for the entire company. This correction function was akin to holding up a mirror to examine the entire organization with its structures, routines and decision-making mechanisms in order to reveal possible blind spots. One of the main areas to be leveraged was management development, as is described below. The Competence Centre for Corporate Management (today: Institute of Corporate Development) of the Bern University of

1 What is detailed here can verify the assertions of the former head of the R&D department. He consistently maintained that innovation is an everyday business, which must be dealt with every day. As he saw it, innovation must happen automatically day in, day out. But at the same time it was interesting to note that his area of responsibility had not in fact produced any really new innovations for some time. As time passed, it also became clear that new ideas were being increasingly rejected with reference to experiences that had been made in the past. The R&D department and its leadership had become locked in their own form of corporate myopia and were no longer capable of embracing new, different or enlarged visions.

Applied Sciences has been observing this process as a participant observer for more than 3 years.

The focus on the innovative management development, which we want to demonstrate, is the classic triangle analysis for organizations: strategy, structure, culture and, added by us, people/human resources and communications [STE 05]. We will show that it is possible to successfully establish innovative management development and, in turn, innovative innovation management by operating on these five levels. Basis for the entire operation is the fundamental principle of reflection. Vigilant organizations must reflect on what they do. This is a prerequisite for innovation and hence sustainability. Our definition casts innovation as a difference that must be perceived as such by an observer and that can be described as new [BRA 05].

The theoretical basis of these descriptions perceives innovations as primarily social innovations. Innovations arise in (communicative) networks, where linking and recombining communications can create something new [HIR 07]. It is no longer sufficient to place responsibility for generating innovations solely in an R&D department as this does not take cognizance of "multidimensional" markets and a "polyphonic" society[2]. There is such a multitude of social trends, new developments and observations that organizations need a "mirror image" of these phenomena, allowing them to be flexible in their ability to observe and act proactively and innovatively where needed, while remaining flexible to react innovatively when the market demands it. Ultimately, it is all about creating an innovation environment, an innovative climate that extends across the entire organization and in which management has a central responsibility to shape this environment. An innovative environment is all about enabling new observations and making them accessible. It is about sharing knowledge and ensuring it is put to practical use. Central to social innovation is ensuring that social relationships are open, and that the contacts arising from them will allow new ideas to be generated and revealed, which in turn will then

2 This of course does not mean that an R&D department is unnecessary or that there is not a case for having several R&D departments. The core competence of developing products would still be anchored in such a department. However, the ability to observe the market and society and the fundamental skills needed to do things differently from how they have been done in the past, must be spread across the entire organization. The R&D department can then access them when it comes to product development. We should never forget that every employee in a company is also a member of society and a customer of other companies. This means they will have ideas how certain products could be designed better or what products still have to be invented. See on the idea of polyphony [AND 03, ROT 10].

generate new/different contact opportunities and new potential for innovation. This potential for innovation goes beyond just product innovations to include organizational and process innovations.

The extent to which innovation is linked to the social sphere can be clearly seen by looking at the "other side", that is at situations where it is impossible to generate innovations or to make them happen. Usually, the social body or the social system decides whether an innovation is adopted or implemented. Path dependence is the buzzword that explains why it is normal to resist change [DIE 11]. Structural factors, normally in the form of anticipated expectations, means it becomes difficult to implement new thinking. Each individual acts and behaves as he/she expects others to act and behave. To take a "tangible" material level, it is easy to observe, for example, that investing in specific machinery will in turn entail the need to invest in specific skills. This will make it unlikely from a structural point of view that an organization, even if the business environment makes it patently evident, will decide against utilizing existing machinery and skills. Those who wish to make an organization innovative must bear two things in mind. They must enable the organization to see just how path dependent it is, and they must ensure that the organization understands innovation to be social innovation. Both of these happened in our case study, and management adopted this observer perspective as a reference point for further action.

Observing path dependency makes it abundantly evident to managers that they must shape management development differently than they have done in the past. This gives rise to the first difference that can be described as innovative at the management development level: first, it is not the creative techniques that are used to generate innovations, but techniques of identifying path dependencies, which, once identified, can be overcome. Second, managers are not seen as people who act, but as parts of the organization and they act as part of the structural organization. This structure is responsible for ensuring that specific (e.g. communication, interaction and decision-making) structures are established, which allow (or not) an observable and communicable social form to be formed, in which innovations can be generated and which can be made available to the organization, that is to the decision-making apparatus.

Using our case study, let us now outline the individual components of innovative management development. This can be used a starting point and as a theoretical framework for observing path dependency and will mirror all subsequent perspectives (strategy, structure, culture, human resources and communication) as a kind of corrective.

3.3.1. *Small event/critical juncture*

Within path dependency theory, it is important to define a so-called "critical juncture" in the context of innovation issues. This presents itself in fact as a paradox: if we want to be innovative, we must be innovative from the very "first" action. Or, in other words: if we want to change things, we must already be changed. In our case, the important "small event", which subsequently proved itself to be a "critical moment", was the redefinition of the HR function, which went beyond traditional HR perspectives and assumed a role beyond its normal mandate. Since there were no predefined structures (i.e. organization's internal path-dependent knowledge) governing how a modern HR function had to be organized, an organizational window of opportunity opened for people to redefine something and to express new ideas, which in turn could be transformed into organizational decisions and procedures. It was of course opportune that HR could depend on the propensity of the founding family to embrace innovation and this facilitated HR's acceptance of its new role. It was also important that this "window of opportunity" happened at a time of crisis. It was clear to management that action was needed, action that was different to what had been done in the past. It took the innovative decision, totally against the mainstream flow, to invest in the company's innovative capabilities at a time of crisis. Two small events, which ex post proved to be a critical juncture, carried within them the nucleus of innovation. And, what was more, embracing innovation as they did kept them, when viewed closely, within their own rhetorical path as an innovative SME. This meant that the innovative novelty could be integrated into the existing organization[3].

---

3 The financial and economic crisis of 2009 was in itself reason enough to change something in principle within the structure. Management analysis showed clearly that, despite a relatively successful market position, follow-up innovations were drying up, business was enacted in the main with the same customers and the time between idea generation and actual implementation was very long. It was also evident that, although innovative ideas were available, these could not be implemented due to the demands of the operative business. On closer examination, it became clear that the company only considered itself innovative due to the myth of its own self-description and the fact that it had prospered in the market to date. It was the crisis that revealed the structural needs for change, and the fact that other forms of communication than self-description were needed and had to be discussed at management level. It was also the structural need for change that allowed the concept of path dependency to be utilized as an analytical and observational tool. This juncture also makes it evident that it was not just any old issue to which the company had to react and question its own structures. To repeat, it was all about innovation, a topic to which the company had always reacted sensitively.

3.3.2. *Strategy*

Despite the fact that innovation had always been regarded as a characteristic structural feature of this family business, management decided during the economic and financial crisis of 2009 to integrate innovation as a long-term strategic element into the company's decision-making bodies. With this move, the company opted for a countercyclical approach: it invested in times of crisis by enhancing innovative capability as the basis for future economic growth (such action was risky: when the decision was taken, it was not clear what the result would be, and whether this in turn would help make the company more innovative, let alone whether meaningful innovations would be generated, which would bring success in the market) [HAU 05]. Part of this strategy was also that all departments had to define for themselves what they could contribute to enhancing the innovative capability of their own area and of the whole company. Management also decided that each business unit had to define how existing processes could be carried out differently. The goal was to create awareness that alternative options were available, and that set ways of doing and deciding things could be done differently if required. Managers were to be evaluated on how they (1) have alternatives to hand and (2) how they drive long-term innovation projects/ideas. The second point was designed, in particular, to prevent pure short-term orientation with its pernicious effect on planning long-term innovation, which is always slightly uncertain [DIE 07].

Central to the strategy was a focus on measures and target groups, which can be represented as follows based on a strategic argument inventory (this, beyond strategic elements, also includes the organizational elements inherent in innovative management development: structure (organization), culture, human resources and communication).

Focus	Measure	Argument
Structure/ organization	Check existing management and leadership tools	The context must be correct to produce innovations. Leadership and management tools are the tools used to control a company. If the innovation aspect is not clearly evident, then appropriate tools are not available to push this topic in a structural sense.

Structure/ organization; personnel	Generating internal cooperation	Innovations can sometimes happen due to coincidences. We want to use such coincidence by bringing people and potential together in a way that has not happened before. We trust that these "new" combinations of people and their ideas will spark new innovations for the company such as would not have happened before.
Structure/ organization	Laboratory as the hothouse of experimentation	Innovation is often premised on trying things. R&D, in particular the laboratory, is the core of our company. Many ideas occur in the laboratory but get lost in day-to-day business. But often it only takes one step in the laboratory to try an idea out – the innovation then takes on physical and plastic form. That is better than "just" writing an idea down on paper.
Personnel, communication, culture	Innovation stars on the "red couch"	There is a face behind each innovation. How have others managed to do something unusual, what obstacles did they need to overcome to get there. The red couch gives us role models to which we can refer. They also provide powerful arguments for our own work with innovations. The red couch is a symbol of our culture of innovation. The goal is that one day we will all have been on the couch.
Personnel, communication, culture	Coaching for managers	If we as a manager want innovative employees, we have got to know how to lead them. Fewer orders, more employee self-organization/self-responsibility. We need managers who can mentor employees on their road to becoming entrepreneurial innovators. Rigid management may be fine in the military – but not with us. We have got to survive on the market every day.

Personnel	Perception-oriented leadership training	If we want to be innovative, we have got to look at things differently. Day-to-day business always means we end up looking at things in the same way. We forget how to use the greatest talent we have: our perception. Perception reveals a lot more than we imagine. Let us start to use this resource and deploy it systematically in the cause of innovation at Rampf. It takes considerable training to first observe rather than act and react in panic. But there is no training we are not willing to undertake.
Structure/ organization	Innovation incentives	Our strategy lays down that we spend 10% of our turnover in training employees. Part of this is ensuring that our employees are given freedom and money when they are developing innovations that came from their own ideas. We are a research-based company, this is our culture. All our employees should be rewarded when they participate in it. Ideas that are developed further will make certain that our success continues into the future.
Communication, structure/ organization	Customer events	Increasingly, innovations are generated together with customers. But that only happens within the right atmosphere. It must be friendly. It must be conducive. We create such an atmosphere. It helps us enter new joint ventures and networks, and these will secure achievement of our financial goals in the long term. In an innovative future, we must stand shoulder to shoulder with our customers.

Note: Author's own illustration.

The semantic distinction we are making is particularly relevant. The strategy was all about enhancing our innovative capabilities, and not about boosting innovation output. This is an important distinction. HR and management understood that innovations cannot be planned. But strategic measures can be enacted, which form a framework that increases the likelihood that innovations can indeed be generated. This is anchoring context management as a strategic element [WIL 01].

3.3.3. *Structure (organization)*

The appeal to be more innovative within an organization is by itself useless. The existing structure is too strong and exerts influence on how staff acts in a day-to-day context. This in turn reinforces the existing structure and the influence it exerts. Together they curtail initiative to keep the organization innovative. This was the reason for introducing company-wide weekly Monday meetings. They were chaired by executives from the top management level. The key topic was innovation. These meetings addressed the question of how what is done every day can be done differently, and what observations were made during the previous week, which may have innovative potential. These meetings should also investigate where, within day-to-day routines, time can be scheduled to examine innovative ideas more closely.

As it was important to break clear of everyday routine, these meetings were held in dedicated innovation corners (Innovation Kitchens), which were specially set up in public spaces within the building. Materials such as Lego bricks, plasticine and building blocks were made available to help participants express themselves. Using means other than speech to represent an idea creates a new form of access and helps others share it.

The Innovation Kitchens (the name is derived from the chemical laboratories from which the company always had to drive its chemical business) also became the location where the company's idea management was repositioned. Dockets were hung at various places within the company where employees could write any ideas they may spontaneously have had. These were then collected in the Innovation Kitchens, and were subsequently referred to and extended/complemented as needed in the Monday meetings. Ideas referred to were merged into the existing project database so that the innovation projects received the same status as classic, established (customer) projects.

Finally, this structure was connected with management and particularly with its top layer, by making each member at this level responsible for one of these innovation corners. A fixed time window was introduced into the weekly management meetings to discuss the results of the Monday meetings and the ideas they had revealed. Top management was also provided with a coach, whose remit was to reflect with them on what had been positive in the new orientation toward innovation, how to deal with employees and managers who were still in denial and what to do to avoid any return to the old patterns of communication and decision-making. Admittedly, this would indeed have been a lot easier than simply holding on to the new ideas emerging, in the case, for instance, that no new ideas were forthcoming on Mondays and staff wanted to return to their work as there was nothing to discuss from their perspective.

Another relevant point should be mentioned. The target objective system was enhanced to include an innovation aspect. Managers were now also assessed on how well they encouraged innovation and the time they gave their staff to be innovative.

Finally, we would like to refer to the last measure enacted to define the new structure, namely the formats used to harness customers as drivers of innovation productions. These are vital, because product innovations in future will no longer come from the cosseted atmosphere within organizations but will instead derive from networks [GAS 12, HIR 07]. Working directly with market participants increases the probability that innovations will be successful and will be accepted by the market. It is extremely important for automotive suppliers to know the innovations on which the car producers or other automotive suppliers are working, as this allows them to anticipate what is going to happen in the market and to develop their own innovations to match. Formats were thus developed in which premium customers were invited to 1 day events, in which the supplier's own existing new products were presented and fields discussed in which a joint innovative production would be useful. These workshops were chaired. The decision-makers wanted to initiate a project structure consisting of members of both the organizations who could start operational work immediately following the first idea session. It was therefore important within the workshops to find out who the key employees were who would be tasked with executing these ideas. Establishing such a format had a strong effect on the internal structure. Incorporating customers in the process meant embracing expectations and commitments that had to be met. Shifting innovative production well away from the boundaries of one's own organization leads, inevitably, to minimizing the danger from blind spots. We cannot get overexcited with our own great ideas if there is a customer who can continually point out that the idea in question has no market relevance for him/her.

3.3.4. *Culture*

At this point, we would like to point out the four cultural design initiatives that were carried out in the context of innovative management development. By culture, first we mean decision premises not open to direct decision processes (they are effective in the social environment of the company and within decision-making itself, without any direct decision having been made on them) [LUH 00]. Second, we mean commonly shared values and artifacts [SCH 92] and third, the way the company deals with the unknown [DOU 89]. To allow innovation to actually gain a hold that transcends mere verbal assertions in the organization's cultural memory, it was vital to involve all staff levels in the debate on this theme. As described previously, management levels could avail themselves of training, reflection and coaching opportunities in order to prepare themselves for what it actually means in day-to-day work to generate operational approaches different to those normally used. How should time management and capacity planning be developed to meet this new requirement? They were also trained to understand what it means to use alternative materials as a way of going beyond verbal expression.

At employee level, new ways of accessing the organization were generated. These transcended what they had been used to up to now and allowed them to see old subject matter in a new way. To facilitate this, a process was developed together with a musician. Staff had to collect and record what they deemed typical day-to-day sounds from within the organization. These sounds then served as the basis for a "corporate sound", which provided a tangible cultural artifact. Today, the basis of this sound can be heard in the company's phone waiting loop, in the corporate film and as elements of a sound landscape on exhibition stands. The sound was presented to all employees at a year-end party, and because employees provided the key elements, they found it easy to identify with it. Of course, the process was more important than the result – why did employees choose certain sounds as typical for the corporation? A key part of the communicative process was coming to grips with what lay behind the choice. What were the images, stories and emotions they wished to evoke? As staff provided what they deemed typical sounds for the organization (more than 70 sound samples were represented, from heavy metal to rock to the classics) and then had to choose what they believed most representative, it became obvious that what was emerging was the very essence [GEE 95] of the company and its culture. Staff could experience within their own company how to perceive what they believed they knew in a new light. This enabled them to draw analogies with their day-to-day work and this, in turn, enhanced the innovative drive. How can we depart from what we believed was the only way of doing something?

The third initiative we would like to refer here is the so-called "red couch". The red couch was conceived as a stage format, in which personalities from other social fields discuss on the subject of innovation. This is billed as a classic "cultural" event and all staff in the company are invited to attend it, as a part of their working day. Three events per year are planned. The personalities come from sport, show business, music and the film industry as well as from other companies. What does innovation mean in the context of sport? How can musicians create something new in a market where it can be assumed that everything that can be composed has already been written? How do other entrepreneurs produce innovations? Staff can then use subsequent plenary discussions to distil an essence of what they have heard and what this can mean for their day-to-day work. Can any of it be seen as a "pattern of action" for their own innovative behavior? This format attempts to maintain the whole topic of innovation in staff consciousness and in the organization's corporate memory while continually re-energizing it with new stories.

The three levels described (strategy/structure, organization and culture) circumscribe the context for human and interpersonal activity. We now want to concentrate on this "human" level that in itself is influenced by these framework conditions, which we have posited as a structure. Successful innovation management can only be effective if there is a structure that encourages people to be innovative. If there are actually people within a company who are particularly creative or innovative, an organizational structure is needed, which allows these characteristics to flourish. Structures should support people and allow them to develop their own innate propensities. They should encourage. Strategies without a supporting structure flanked with concrete initiatives for actual implementation are never successful. Structures that are not stiffened with cultural aspects (beliefs, values and attitudes) that promote actual use of the structures are hollow. These structures are purely verbal and cannot be successful on the actual operative level. Just talking about an innovative corporate culture without any form of strategic underpinning is a purely rhetorical exercise – it has nothing to do with actual implementation. It needs to be linked to the structural level if innovative management is to become the cornerstone for innovative innovation management.

3.3.5. *People/personnel*

The level of people and personnel was processed in our case study chiefly via management and personnel development seminars and workshops that we have mentioned. Coaching is a very important element for facilitating the "innovative impregnation" of the company. Coaching is all about leading managers and staff to the boundaries and fault lines of their respective areas, allowing them to experience the "personal and physical level of consciousness" when understanding their respective roles at the flashpoint between individuals, their roles/functions and the organization[4]. And there are many roles in innovation management, as we will discuss. We are working with people here, tracking them as they move within and become acquainted with structures. Initiatives were taken to identify missing skills and correct them. A key question that emerges on the personal level is as follows: how can day-to-day work be maintained to meet quality and scheduling standards when at the same time the organization is propagating the call to embrace innovation and thus deviate in part from these very norms and standards? How can innovation and the innovative quest be given priority, if a customer-driven producing organization (such as is the case here) can always argue that what is actually happening in the market has ultimate priority, and that innovation should be a peripheral activity at the operative level?

Another way to directly influence people is the reshaping of the goal achievement incentive system as already discussed under the topic of structure. This integrates into evaluation schemes how well managers have embraced (longer term) innovative projects and what they are doing to encourage their staff to act innovatively. Such a structure is designed to have a direct influence on how people act.

Finally, it was important when new appointments were to be made at all management levels, as was the case, for instance, when the management of R&D and the laboratories was restructured in its entirety, that candidates for the positions had the correct profile in terms of their ability to promote innovative behavior.

---

4 For more on this his tension triangle and coaching see [BAC 04].

At the personal level as well, we would like to point to a "small event", which in retrospect proved to be a "critical juncture": the process of embracing innovation and making it a priority topic within the company would not have happened had there not been the right CEO and a founding family. What is needed are people who can stick to decisions even when many in management cannot fully embrace the vision due to the demands of day-to-day business. Seen this way, the right people were in the right positions, could make the right decisions and could install the right structures [LUH 00]. What is needed are people who stand fully behind the topic in question, in this case innovation. We need organizational access points that stay reliable over time and across all communication channels.

3.3.6. *Communication*

If we go on to consider communication as a final component of our innovation management model, we must realize that it plays a seminal role. From a systematic and constructivistic organizational perspective, it becomes a reference point for the innovative quest. In our case study, innovation or innovative capability became the red thread around which all initiatives could be oriented within the organization. For example, innovation is a central argument in the strategy we have mentioned earlier: innovations are needed in order to be successful in markets in the future. Strategy does not have an existence of its own, it must be communicated. And innovation was always the focal point of such communication in our case study.

Even in the case of communication directed at management, innovation was always a key argument in creating situations in which staff were encouraged to make use of their innovative potential.

Cultural communication needs to take every opportunity of reminding the organization that its goal and guiding values are rooted in the quest for innovation. The "red couch" was cited as an example of such a communicative awareness initiative. Space is needed in which coaching and management and employee development can take place, in which communication can take place and in which a common awareness of innovation can be conveyed. We can see from our case that communication and the correct platforms had a central role in ensuring that the organization and its staff could construe the true meaning of innovation and see it as a means of securing success into the future.

3.4. Reflection as a key to innovative management development

The truly innovative feature of this management development approach is the way in which reflection is given a structural anchor within the process as it relates to the management level and management development. Managers are encouraged to reflect on the results their initiatives have produced, to reflect on how they act as managers, to reflect on the whole innovation strategy and, if need be, to adapt it accordingly; they are to reflect on alternative approaches for exploiting the organization; and they are to reflect on just how much their own actions are contributing to the innovative culture of the company. They must reflect on how to be innovative in their own approach to innovations.

This reflection process is supported from two sides: first by the HR function, which aims to boost the company's innovative competence and reveal the company's blind spots, and second by external consultants. Consultants' reflections on the whole innovation process must prevent the organization, management and HR from falling prey to a blind spot. These consultants are tasked with introducing further innovative ideas from their perspective for company development, besides observing the entire system from outside as a way of pointing out any redundant routines and blind spots that may have pervaded innovative production processes. Integrating this as a function and structure into innovation management is the only way of ensuring innovation management is in a constant state of reassessment, and only when that is the case it can truly be called innovation management. Only when there are sufficient opportunities to reassess existing conditions and hence also the whole innovative management process, and only when management can decide such issues, are we talking about an innovative innovation management, which, as we have described, must initially build off an innovative management development process. It is not "just" about creating a routine for an innovation process, it is also about creating a routine for permanent reassessment of the innovation process. This may be a subtle distinction, but it is a crucial one when the task is to bring constant innovative momentum to the theory and practice of innovation management. (For the limits and possibilities of creating a routine for innovation and innovation processes, see e.g. [KOC 09].)

3.5. Further research issues

A central question is how a continuous, positive reassessment of the organization can be institutionalized and fostered, with the aim of creating an ongoing awareness for the whole topic of innovation. It is crucial that relapses into old patterns and paths typical of organizations not aligned for innovation be avoided. As referenced earlier, blind spots within an organization are a problem as these point up perception gaps. But only if a company is aware and capable of perceiving problems, dangers and also opportunities, can it then result in (positive) change for the company.

Against this background, ongoing communication on the relevance of innovations for company survival and future growth is of particular importance. Analogous to Luhmann's system theory, the individual (sub-) systems must communicate with each other about a topic, in this case the whole area of innovation. Only if enacted on a topic, here innovation, can that topic exist for each (sub-)system. This illustrates the great onus that must be laid on communication. The actual mode of communication is especially interesting from a research standpoint. It is not "what" is communicated, which is decisive, it is "how" it is communicated. Multi- and trans-disciplinary research could help secure an integrated view of the whole subject context. It might be feasible to conduct research from a business, artistic and linguistic perspective. Research questions in this context might be the following.

How can art be used as an intermediary in innovation processes and as a way of boosting innovative capacity in companies? What tools from the whole process of artistic creation can be applied to the business sector? How can features of art, such as the ability to reflect, analyze and reveal organizational deficiencies be applied in organizational contexts? A further investigation of the integration of science and art as a research topic could include linguistic research. How and using what kind of language can the individual (sub) systems and protagonists communicate with each other? It certainly would be interesting to investigate what processes are involved in the process of reaching a common agreement and how these can be positively influenced.

Research topics in the context of innovation are many and varied from a business administration point of view and a number of results, examples and experiences are available. However, building on the various questions posed above, there is plenty of potential to broaden the research view of the whole topic via integration and cooperation with other research disciplines. Research that targets these interfaces often delivers interesting and useful results for theory and practice.

3.6. Bibliography

[AES 09] AESCHBACHER R., *Maximale Innovation: Durch Management by Conversation*, Rüegger, Zürich, 2009.

[AND 03] ANDERSEN N.A., "Polyphonic organizations", in BAKKEN T., HERNES T. (eds), *Autopoetic Organization Theory: Drawing on Niklas Luhmann's Social System Perspective*, Copenhagen Business School Press, Copenhagen, pp. 151–182, 2003.

[ARR 76] ARREGER K., *Innovation in sozialen Systemen 1. Einführung in die Innovations theorie der Organisation*, Paul Haupt, Bern, 1976.

[ATU 12] ATUAHENE-GIMA K., "What is innovation?", *African Business*, pp. 66–67, October 2012.

[BAC 04] BACKHAUSEN W., THOMMEN J.-P., *Coaching*, Gabler, Wiesbaden, 2004.

[BAR 09] BAREGHEH A., ROWLEY J., SAMBROOK S., "Towards a multidisciplinary definition of innovation", *Management Decision*, vol. 47, no. 8, pp. 1323–1339, 2009.

[BRA 05] BRAUN-THÜRMANN H., *Innovation*, Transcript Verlag, Bielefeld, 2005.

[CAR 99] CARR N.G., "Forethought: visualizing innovation", *Harvard Business Review*, vol. 77, no. 5, p. 16, 1999.

[DAM 91] DAMANPOUR F., "Organizational innovation – a meta-analysis of effects of determinants and moderators", *Academy of Management Journal*, vol. 34, no. 3, pp. 555–590, 1991.

[DAM 96] DAMANPOUR F., "Organizational complexity and innovation: developing and testing multiple contingency models", *Management Science*, vol. 42, no. 5, pp. 693–716, 1996.

[DEV 98] DE VRIES M., "Die paradoxie der innovation", in HEIDELOFF F., RADEL T. (eds), *Organisation von Innovation: Strukturen, Prozesse, Interventionen*, Hampp, Munich, pp. 75–87, 1998.

[DIE 07] DIEVERNICH F., *Pfadabhängigkeit im Management*, Kohlhammer Verlag, Stuttgart, 2007.

[DIE 11] DIEVERNICH F., "Pfadabhängigkeitstheoretische Beiträge zur Zukunftsgestaltung", in TIBERIUS V.A. (ed.), *Zukunftsgenese: Theorien des zukünftigen Wandels*, Springer VS, Wiesbaden, pp. 57–72, 2011.

[DOU 89] DOUGLAS M., "A typology of cultures", in HALLER M., HOFFMANN-NOWOTNY H.-J., ZAPF W. (eds), *Kultur und Gesellschaft*, Campus, Frankfurt am Main, pp. 85–97, 1989.

[GAS 12] GASSMANN O., FRISIKE S., *33 Erfolgsprinzipien der Innovation*, Hanser, Munich, 2012.

[GEE 95] GEERTZ C., *Dichte Beschreibungen*, Suhrkamp, Frankfurt am Main, 1995.

[GOS 05] GOSWAMI S., MATHEW M., "Definition of innovation revisited: an empirical study on Indian information technology industry", *International Journal of Innovation Management*, vol. 9, no. 3, pp. 371–383, 2005.

[GRA 08] GRASSHOFF G., "Innovation: Begriffe und Thesen", in GRASSHOFF G., SCHWINGES R. (eds), *Innovationskultur: Von der Wissenschaft zum Produkt*, vdf Hochschulverlag, Zürich, pp.13–31, 2008.

[HAU 05] HAUSCHILDT J., SALOMO S., "Je innovativer, desto erfolgreicher?: Eine kritische Analyse des Zusammenhangs zwischen Innovationsgrad und Innovationserfolg", *Journal für Betriebswirtschaft*, vol. 55, no. 1, pp. 3–20, 2005.

[HEI 98] HEIDELOFF F., RADEL T., "Innovation in Organisationen: Ein Eindruck vom Stand der Forschung", in HEIDELOFF F., RADEL T. (eds), *Organisation von Innovation: Strukturen, Prozesse, Interventionen*, Hampp, Munich, pp. 7–39, 1998.

[HIR 07] HIRSCH-KREINSEN H., "Genese und Wandel von Innovationsnetzwerken – Produktionstechnische Entwicklung", in BERGHOFF H., SYDOW J. (eds), *Unternehmerische Netzwerke*, Kohlhammer, Stuttgart, pp. 119–141, 2007.

[KOC 09] KOCH J., "Innovation, organisationale Routinen und strategische Pfade", *Zeitschrift für Management*, vol. 4, no. 3, pp. 189–208, 2009.

[LUH 00] LUHMANN N., *Organisation und Entscheidung*, Westdeutscher Verlag, Opladen, 2000.

[OEC 05] OECD, *Oslo Manual: Guidelines for Collecting and Interpreting Innovation Data*, 3rd ed., OECD, Paris, 2005.

[ROT 10] ROTH S., SCHREIBER L., WETZEL R. (eds), *Organisation multimedial: Zum polyphonen Programm der nächsten Organisation*, Carl Auer Verlag, Heidelberg, 2010.

[SAF 07] SAFFER D., "The cult of innovation", *Business Week*, p. 89, 5 March, 2007.

[SCH 92] SCHEIN E., *Organizational Culture and Leadership*, John Wiley & Sons, 1992.

[SCH 34] SCHUMPETER J.A., Theorie der wirtschaftlichen Entwicklung. Eine Untersuchung über Unternehmergewinn, Kapital, Kredit, Zins und den Konjunkturzyklus. Berlin, Duncker & Humblot, 1997; 9. Aufl. unveränderter Nachdruck der 4. Aufl. von 1934.

[SCH 42] SCHUMPETER J.A., *Capitalism, Socialism, and Democracy*, Harper & Brothers, New York, 1942.

[SIG 06] SIGUAW J.A., SIMPSON P.M., ENZ C.A., "Conceptualizing innovation orientation: a framework for study and integration of innovation research", *Journal of Product Innovation Management*, vol. 23, pp. 556–574, 2006.

[SIM 09] SIMON H., *Hidden Champions of the Twenty-First Century: The Success Strategies of Unknown World Market Leaders*, Springer, New York, 2009.

[STE 05] STEINMANN H., SCHREYÖGG G., KOCH J., *Management: Grundlagen der Unternehmensführung*, Gabler, Wiesbaden, 2005.

[STE 07] STERN T., JABERG H., *Erfolgreiches Innovationsmanagement: Erfolgsfaktoren – Grundmuster – Fallbeispiele*, Gabler, Wiesbaden, 2007.

[SUB 11] SUBRAHMANYA, M.H.B., "Technological innovations and firm performance of manufacturing SMEs: determinants and outcomes", *ASCI Journal of Management*, vol. 41, no. 1, pp. 109–122, 2011.

[THO 06] THORN N., MÜLLER R.C., "Innovations management in KMU", in BRUCH H., KRUMMAKER S., VOGEL B. (eds), *Leadership: Best Practices and Trends*, Gabler, Wiesbaden, pp. 251–264, 2006.

[TUS 97] TUSHMAN M.L., O'REILLY C.A. III, *Winning through Innovation: A Practical Guide to Leading Organizational Change and Renewal*, Harvard Business School Press, Boston, MA, 1997.

[VOL 10a] VOLKMANN C.K., TOKARSKI K.O., GRUENHAGEN M., *Entrepreneurship in a European Perspective*, Gabler, Wiesbaden, 2010.

[VOL 10b] VOLKMANN C.K., TOKARSKI K.O., "Soziale Innovationen und Social Entrepreneurship", in BAUMANN W., BRAUKMANN U., MATTHES W. (eds), *Innovation und Internationalisierung*, Gabler, Wiesbaden, pp. 151–170, 2010.

[VON 88] VON HIPPEL E., *The Sources of Innovation*, Oxford University Press, Oxford, 1988.

[WIL 01] WILLKE H., *Systemisches Wissensmanagement*, UTB, Stuttgart, 2001.

Chapter 4

Innovative Product Design and Development through Online Customization

This chapter investigates the appropriateness of virtual reality and online customization through e-product development (E-PD), which allows the designers to utilize the Internet's communication with the customers along with great speed and global reach. The chapter informs about the state of play patterns that assist manufacturing firms in employing advancements in information technology (IT) and the Internet for product development and the successful transition en route towards mass customization. It identifies pertained opportunities, possible barriers for the E-PD establishment embracing online customization and the virtual reality tools for potential enterprise manufacturing.

The chapter introduces a conceptual framework for reconfigurable product development and a systematic E-PD loop in order to develop the future customer-oriented perspective and interactions with the customers through the E-PD enablers. The research work focuses on the Indian furniture industry, which could benefit from online customization features to reflect consumer needs, and in turn could implement mass-customization production more efficiently. In particular, the E-PD feasibility and effectiveness is analyzed for the office furniture segment of India, the small and medium

---

Chapter written by M. Reza ABDI and Vipin KHANNA.

enterprises (SMEs), through conducting a questionnaire survey and semi-structured interviews. Accordingly, the readiness of the office furniture industry is examined for adoption of a suitable E-PD prototype in order to facilitate understanding of the current Indian market and move toward responsive production.

4.1. Introduction

With the arrival of the 21st Century, manufacturing industry has been facing increasing challenges from global markets. Today's manufacturers have to cope with problems arising from a greater product mix, smaller batch sizes, shorter lead-times and shorter product lifecycles. Changes in products and demands can happen faster than ever before. Therefore, manufacturing organizations should provide sufficient flexibility to produce a variety of products on the same system. The need to respond rapidly to demand changes for sustainable competitiveness in global business environment creates a need for new strategies and tools for customized product design and development.

E-commerce and mass-customization will emerge as a primary means of transforming manufacturing in terms of core activities such as design, resource planning, factory automation and supply chain management in the coming decade and beyond [JIA 06]. Accordingly, many processes can be re-engineered to utilize Internet technologies, resulting in substantial cost savings and reduction in lead times. According to Higgins [HIG 01], many online approaches are in the process of improving product development processes throughout a variety of industries. These approaches, called E-PD, allow the designers to utilize the Internet's communication speed and global reach. E-PD has speedily evolved from simple email apportioning of design iterations between teams to real-time alliances between geographically separated design teams. As 80% of a product's final cost and time to market is decided in the design process [AND 90, BEL 09], in the short run, the Internet offers enormous potential for creating value by reducing this time. Many industries from automotive to software firms are utilizing E-PD to reduce the cost and time used in product development. E-PD makes it possible to incorporate the voice of the consumers into all product design and development phases, promoting market approval of products beyond the 50% success rate broadly experienced [HIG 01]. Further than the cost and time economies, E-PD, through faster communication, can assist in diminishing transactional costs and augment the scope of collaborations.

SMEs confront a regular plight concerning the issue on how to effectively place themselves against both larger, more financially stable rivals and smaller, nimble niche players. To be successful, the SMEs need to project an image of size, substance and steadiness on an exceptionally limited budget. However they have been fast enough to hunt unique niche chances that may go unnoticed by the competition. In today's business environment, on-going systemic adaptation within businesses is obligatory. Customized solutions are useful for both consumer acquirement and preservation and can spawn successful, long-term relationships [ANS 03]. The Internet can improve the collaboration of people involved in the product development process in SMEs by intensifying the speed and quality of product testing and the validation and amelioration efficiency and effectiveness of manufacturing planning. The Internet has a positive impact on product management activities, which includes the faster detection of user requirements and superior product customization [AVL 00], and can be accessed from a broader range of locations with a wider span of ideas and thoughts for product development [OZE 03]. Individuals, for instance customers, experts and salespeople, in different parts of the world can participate in the concept generation process and hence augment the ability of the firm to yield a wide range of ideas from a broad range of idea sources. Thus, numerous design ameliorations can be completed in the electronic prototype prior to a much more costly physical prototype and they can be tested in a much shorter period of time. As animated prototypes can be created and communicated with all the concerned parties in a short period of time, the Internet can considerably increase the speed of testing and validation. In addition, as creating an online prototype is usually much more economical than building a physical prototype, the Internet can also reduce the cost of testing and validation. Because of the reduced costs, firms can carry out the various tests early in the product development process and can effortlessly repeat them. In view of the fact that repeated testing is coupled with more accuracy, the Internet can also boost the quality of the tests [OZE 03].

Many researchers have already demonstrated how the Internet can help businesses in their core activities such as marketing and sales [SAM 97], manufacturing [OZE 03] and purchasing [STU 97]. However, very few research works have addressed the impacts on E-PD in an Asian country [KAW 11], and to the best of the authors' knowledge no research work has been published addressing E-PD in the furniture industry of India. This chapter is intended to improve understanding of the E-PD concepts, process and implementation. The following sections investigate the feasibility of E-PD implementation, embracing online customization and virtual reality tools for the SMEs of the

Indian office furniture sector, in order to improve customer satisfaction and enhance operational requirements for competency against the big firms.

4.2. Mass customization and reconfigurable products for E-PD

What distinguishes mass-customization from simple customization is the integration of computer-based information systems with advanced production systems such as flexible manufacturing and reconfigurable manufacturing. The market and manufacturing are then linked to provide unique products for each customer [BAR 03]. Reconfigurable manufacturing is designed to employ the necessary flexibility (no more no less) of manufacturing facilities that require the right capacity and functionality in order to quickly create products that meet customer expectations for customization, variety, low cost and high quality. For reconfigurable manufacturing, products need to be grouped within families according to their operational similarities [ABD 04]. Standard products with fixed configurations, that is, only some predefined types, are offered. Customers merely can choose between these types, but no changes or extensions (at least at the system unit) are possible. Customized products must be fully configurable for the designated variety. In this case, the customer specifies which components he/she wants to get from which supplier or at least the options of the components he/she wants to get. The manufacturer tests whether the requested configuration is technically feasible and calculates the price. Because of the ability to combine many different components – again obtainable from several alternative suppliers – an incredibly large number of possible final items are given.

With shifting the market toward the mass customization, product lifecycles and development cycles are invariably reduced. Therefore, modularity becomes the key to accomplishing low-cost customization [DUR 00] by reusing parts and components and the efficient use of manufacturing resources.

Mass-customization extends the opportunity to comprehend and acquire potential market niches and also aids in the development of technical competences to endure diverse demands of target customers. Therefore, we loom toward mass-customization from the product development perspective. The basic strategy for it is to embrace customers in the product development lifecycle by proactively associating consumer's requirements with the competencies of a company.

The Internet makes mass-customization eminently possible. Customization is possible in part because of the interactivity afforded by the

Web [ANS 03]. Nowadays, economy is backed by information businesses. Information has the benefit of being easy to distinguish, customize, personalize and dispatch over an electronic network at unbelievable speed [KOT 09]. Customization begins from the need for specialization and may be a psychological effect, not a technological issue. Specialization extends to personalization or customization where the individual customers can tailor-make products to their own specifications and then order them. This translates into greater degrees of customer gratification and loyalty [VIA 04].

Miranda and Kónya [MIR 07] emphasized that due to greater media exposure, education and travel, consumers are becoming more sophisticated and do not want to be stereotyped. Cova *et al.* [COV 07] defined them as a new breed of "new economic order consumers"; consumers who take pride in their own identity and are driven by their strong beliefs. Miranda and Kónya [MIR 07] concluded that modern consumers demand to take part in the getup of the product of their choice and also want to be "involved" in the product composition. The Internet especially renders a medium to make one-to-one personalization practical. Furthermore, the influential baby boomer generation of consumers has a strong sense of distinctiveness that leads them to seek customization and personalized products [COV 07].

Song *et al.* [SON 07] notified that product use such as furniture consumption should integrate pleasure and fun into what they call an "experiential view" of consumption. Virtual experience connects users in effective product erudition and gratified experience. People spend 50% more time on a Website that offers customization of products and interactive three-dimensional (3D) images of the same [HUR 00].

Today's customers are taking more personalized initiative in ascertaining what and how to buy. They log onto the Internet, refer to information and evaluations of goods offers and product reviews and in many cases desire to design the product they require. Kotler and Keller [KOT 09] notified that customization combines operationally driven mass-customization with customized marketing in a way that empowers consumers to design the product and service offering of their choice. The manufacturer needs no prior information about the customer in the same way as the business needs no information about the manufacturing setup. The business firm furnishes a platform and tools and "rents" out to customers the means to design their own goods. Some online companies today are extending customers a choice board, an interactive online system that permits each client to design their own product by selecting from a menu of attributes, components, prices and delivery options. The client's choices send out signals to the producer's manufacturing system that launches the process of procurement, assembly and delivery.

Hoeffler *et al.* [HOE 06] reports that consumer preferences are stable and generally change in a predictable fashion. Preference stability is decisive for personalization as the customer makes future choices on the basis of it. The fact that the Internet permits consumers to gain knowledge from product experiences and shared information of other similar clients would also augment preference stability, allowing personalization [IYE 03].

Syam *et al.* [SYA 08] mentioned that the risk of product customization is that the user may realize after designing their "ideal" product that their real preferences match more closely standardized goods. As the pace of technology development and its potential to improve life and work processes are altering, the market is characterized by dramatic adjustments in market offerings, resulting in even veteran purchasers possibly not knowing – or may being overconfident in – their preferences and choices. One of the ways to diminish preference vagueness is by facilitating buyers with interactive tools that allow them to visualize and experience customized products prior to purchase [LUR 07].

According to the findings of Goldsmith and Freiden [GOL 04], one precise hurdle that lies in the perception of mass-customized goods is that they cost more. The actual price charged for the mass customized goods must be equal to or even lower than the non-mass-customized commodities, or the thought that they cost more could be made a positive selling point. One way to make it feasible is through lowered production costs and inventory.

According to Stadtler and Kilger [STA 03], the degree of customization varies between the two following extremes:

1) Standard products with fixed configurations, that is only some predefined types are offered. Customers can merely choose between these types, but no changes or extensions (at least at the system unit) are possible.

2) Customized products that are completely configurable.

4.2.1. *Innovative reconfigurable product development*

Agile manufacturing is the ability to use a flexible, diverse production environment to quickly create products that meet or exceed customer expectations for customization, variety, low cost and high quality according to Goldsmith and Freiden [GOL 04]. To facilitate the selection of the products in the production range, and transfer customer requirements and manufacturing capacity and functionality, a permanent reconfiguration link between market and manufacturing is required [ABD 03]. Furthermore, conventional customer-oriented product development methodologies are concentrated on the current needs of the current customers. On the contrary, the reconfigurable product planning and standardization could expand the focus of all three dimensions, that is customer, time and product specifications. Figure 4.1 depicts an expandable 3D perspective for reconfigurable product development, whereas the small solid cube surrounded by the larger cube representing reconfigurable product development depicts the classical customer-orientated perspective.

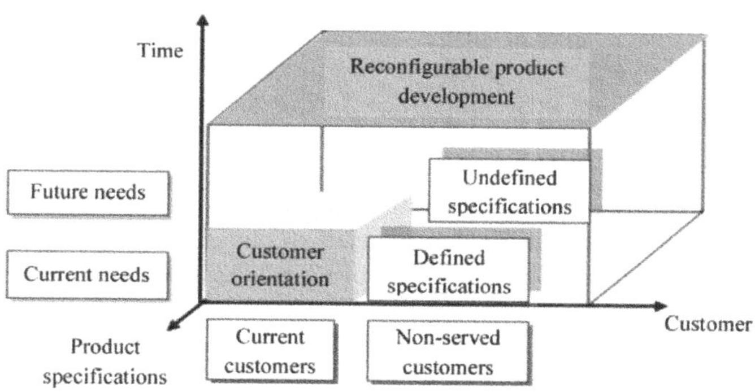

Figure 4.1. *Product development based on conventional customer orientation versus reconfigurable product development*

4.2.2. *Conceptual systematic loop of E-PD*

With the rapid growth of the Internet, e-commerce and mass-customization conjectures, typical companies, which pursue design, manufacturing and logistics processes, will become rare [HEL 02]. Therefore, a platform enabling networking among customers, product designers and suppliers seems to be essential. Accordingly, a conceptual systematic loop for E-PD and virtual reality, as shown in Figure 4.2, can be derived through three pillars: the integrated product design and development in the product lifecycle, mass customization of the product in the production/supply range, and supply chain management and e-commerce infrastructure. As it is a dynamic system with a feedback loop, for every new product or product customization, the process has to "go around the loop". By employing the online customization via E-PD on the doctrine of mass customization and personalization, a greater level of customer satisfaction can be achieved. Consumers can take advantage of self-guided selection and (re)configuration tools to choose products and options. The business is capable of trading merchandize with distributors and/or directly with consumers via the supply chain and e-commerce. Products are designed in consensus by the customers themselves, and thus customer requirements are captured directly through their preferred electronic design.

Involving the customers in designing could help the firms to gain insight into the decision-making process and more importantly feedback from the customers regarding their choices. It is crucial to provide customer feedback so that firms can manufacture furniture that will be acceptable to buyers. The apprehension of a customer's decision processes and impediments that customers face during the processes is decisive to the success of the industry.

In addition, realizing which designs are refused by customers could add enormous value to companies since it reduces costs by minimizing inventory and distribution clogs. Informed design and management are vital for the office furniture industry to survive and succeed. In conjunction with online customization, web-based virtual reality can also offer indirect but realistic visual feels for accessing office furniture to visualize if it matches other furniture items and decoration.

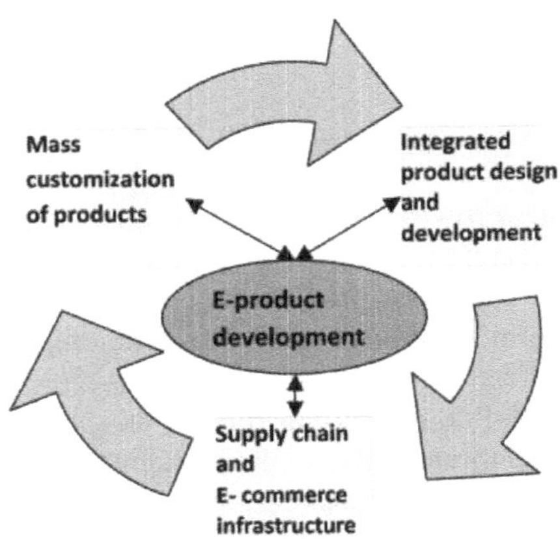

Figure 4.2. *Conceptual model of E-PD enabled mass-customization*

4.3. The empirical research design

The research objective is to analyze the appropriateness of virtual reality and online customization in E-PD, to be employed in the office furniture segment of India. The furniture industry could benefit from using distinguishing features derived from virtual reality and online customization in order to understand consumer needs and wants, and in turn could implement mass-customization more efficiently. The data were collected to measure and analyze the readiness of the office furniture industry to adopt online customization. To have an understanding of the Indian market, an overview of the current office furniture market will also be provided.

Figure 4.3 illustrates the research design employed for preparing and carrying out the research project. The proposed research methodology explains gathering primary and secondary data. This section also presents the data accumulation method, sampling techniques used and the potential constraints of the data compiled. The project begins with the review and analysis of the current literature on new technologies, such as online customization and virtual reality, which currently exist in the market. Research questions are then designed by employing the mass-customization dimensions and tailoring them to the specific requirements of the office furniture industry. The gathered data are used to present the analyzed outcomes through a deductive approach as the way to arrive or not at a new finding. To discover the readiness of the infrastructure for implementing online customization and virtual reality in the office furniture industry, internal and external factors, along with the stakeholder's perceptions, are studied. Thus, the research strategies emphasize explanatory investigation.

To collect the data, as part of the data triangulation process, a three-stage process was applied as follows. The first stage required an extensive exploration of the literature to establish the extent and limits of E-PD for the mass-customization paradigm keeping in focus online customization and virtual reality, in the context of small office furniture business organizations in India. Despite the obvious interest in this area, little published research exists which addresses these issues in a way that allows a holistic view. To overcome this, the second stage of data gathering was commenced, which comprised a series of informal telephonic interviews with the architects and the office furniture dealers in North India. The third stage involved the utilization of an electronic questionnaire, which was distributed and returned through email.

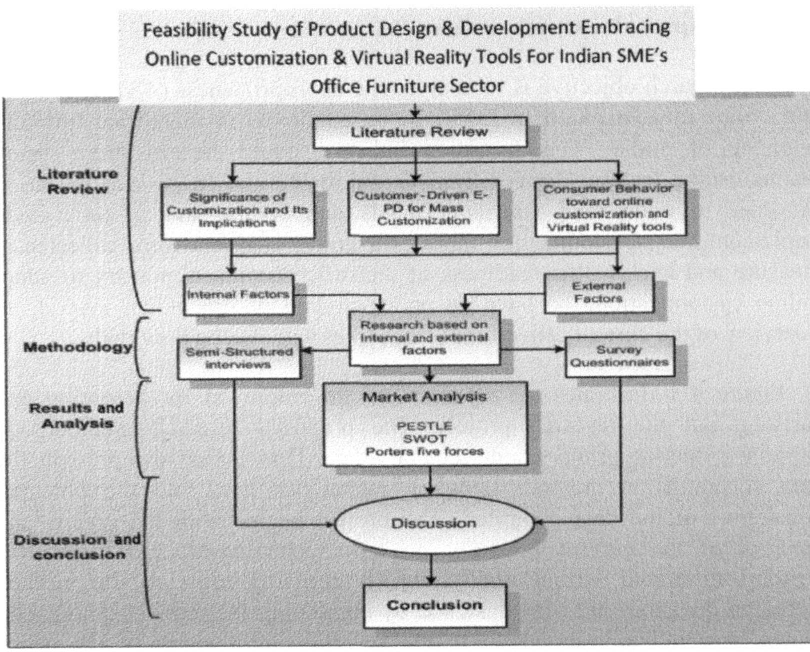

Figure 4.3. *The research design and methodology*

Figure 4.4 highlights four quarters as assortments of data source classifications including primary data, secondary data, internal data and external data as inputs to the various factors (shown in boxes) influencing the feasibility study (central box). The quarters are the assortments under which various factors are assessed for appropriateness and acceptability of the E-PD in office furniture firms. Accordingly, quarter 1 presents the status in which the primary data collected from the internal data sources affect the strengths and weaknesses of the firms; quarter 2 presents the status in which the secondary data collected from the internal data sources affect the technical and environmental factors; quarter 3 presents the status in which the primary data collected from the external data sources affect the competitors, end consumers and intermediaries/agents for financial capabilities and decision-making, and quarter 4 presents the status in which the secondary data collected from the external data sources affect macrocosmic factors of India and industrial analysis (import/export of furniture).

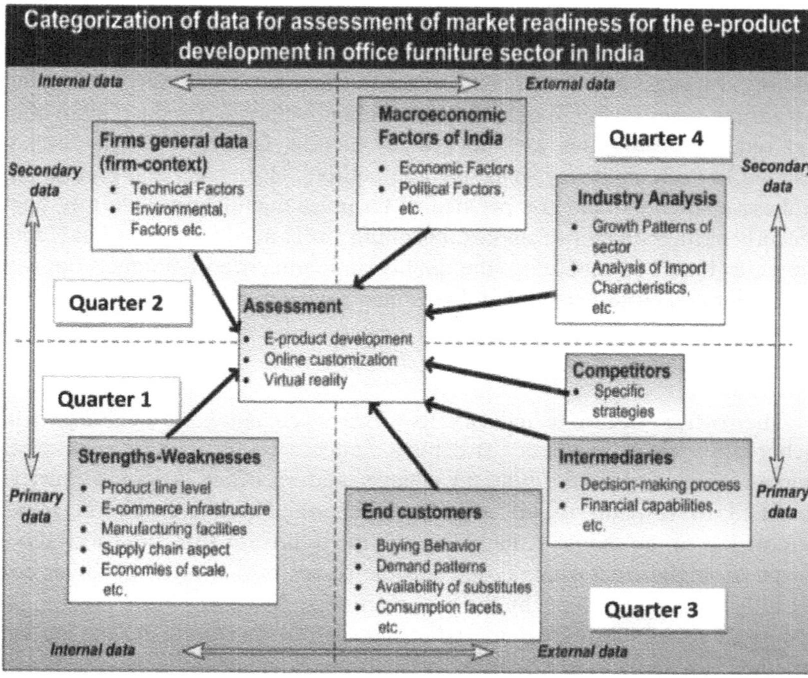

Figure 4.4. *Data categorization for e-product development and market assessment*

4.4. Case of Indian office furniture sector

The office furniture industry in India constitutes the companies primarily engaged in manufacturing furniture designed for office use such as chairs, desks, workstations, office partitions and panel systems, as well as filing and storage systems. Suppliers of parts, frames, components and fixtures are also included in this sector. Among all the types of furniture used, the office furniture segment is the one that boasts the most significant companies, both from the standpoint of size and the technological innovation of production. These industries have a wide range of clients in the private as well as public sector in India. Almost any business necessitating office furniture is a potential client of this industry.

The office furniture sector is subjugated by SMEs. This is both a point of strength for the Indian manufacturers, as these firms are generally more pliable and quick in accommodating market changes, and a point of weakness, as they are more likely to be deficient in investment capabilities. Rapidly varying and extremely competitive market conditions are putting increasing pressure on Indian furniture manufacturers. Product innovation and reduced lead times are the critical success factors for businesses to respond to such challenges and stay competitive. Horizontal and preliminary to these elements is the competence of furniture manufacturers to rely upon decently skilled human resources. Furniture firms need to pursue innovation strategies based on creativity, the quality and eminence of products, as well as ameliorated customer service. Investments need to be made in the fields of automation of production processes, design and creation, advertising and communication.

The office furniture industry is always changing – evolving with technology shifts-business structures alter, which leads workplace demographics and work styles to change, and so office furniture must be designed to respond to all of these changes. There are several trends percolating in the industry, including the move toward smaller workstations driven by higher real estate costs, and the recent proliferation of lower cost yet fully featured products by the manufacturers. As shown in Figure 4.5, the import of office furniture into India has sharply increased during the last decade. The Indian office furniture industry is contending to survive as imports from Italy, China and other developed countries are becoming significantly higher [IND 07b].

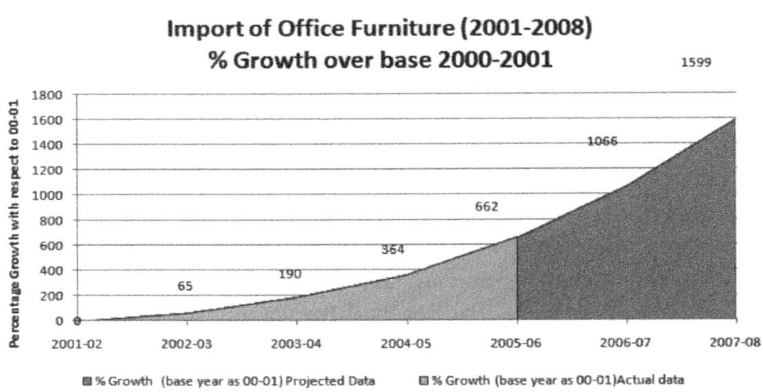

Figure 4.5. *Import of office furniture (2001–2008 [IND 08a])*

Since the Indian office furniture industry cannot compete with imports on cost only, the industry must become more proficient at responding to customer's tastes and personal requirements by the incorporation of efficient marketing and supply chain management. Except for a few big brands, the industry does not altogether understand consumers' furniture buying behavior [MCK 07]. Consequently, the industry is facing difficulty in exploring the market to realize what the requirements of office furniture users are. Moreover, nowadays customers want to try out office layouts, individual spaces and privacy levels, and cooperative work areas. They wish for office furniture planned to their needs – including day-to-day alterations and personalization as well as project-based workplace designing – and to respond to the continually evolving workplace requirements, regardless of what they might be. The vast, dynamic nature of the Indian economy and the strong increase in the purchasing power of its population have enabled India to become one of the countries most interested in "Made in Italy" products. However, the presence of Italian brands and products, especially in the furniture and accessories sector, is very limited and essentially still a niche market. These actions have brought about tangible results: imports from Italy moved from €7.2 million in 2005 to €16.5 million in 2010, making the country the first commercial European partner in the sector [IND 11].

4.4.1. *Market segmentation*

The office environment is an evolutionary march. Almost 20 years ago, rows of desks started to give way to the rise of furniture systems – a blend of varied height panels, protruded work surfaces and incorporated storage and work tools that created extremely flexible and efficient workstations for the rapidly growing white collar manpower. Consumers are drifting away from the "cubicle" concept, that is, high panels distinguishing workstations. Nowadays, lower cloth-covered or glass partitions are used in angular configurations, with less space amidst workstations. Currently, the office furniture market is inclined toward the incorporation of novel materials to develop lighter, more mobile finished goods. New office furniture designs, particularly seats, stress ergonomics; that is balance between esthetics and physical comfort [IND 07a].

Customers in the office furniture market are divided into two prominent classes: "A" and "B". Class "A" comprises multinationals and prominent local corporations that buy equipment from big local firms. Segment "B" includes small firms and individuals that acquire office furniture from importers and small manufacturers. As already mentioned, category "A" embraces the consumers with great purchasing power, for instance multinationals and large business houses. This segment of the market has a preference for quality over any other consideration when acquiring office furniture. These users employ the services of an architect or an interior designer, who not only proposes what office furniture to acquire, but also provides worthful suggestions on other matters, such as lighting, circulation and décor and acts as mediator between the manufacturer and the consumers. The decision-making is a multifaceted process that requires the involvement of the consumer, the architect and the building contractor to varying degrees, devolving on the project. Manufacturers need to target all of them; acquiring approval from one does not inevitably assure the support of all. Businesses in the financial sphere, in particular banks and public service corporations (for instance telephone, airlines and IT companies), are most likely to invest in new equipment. Interestingly, within this market segment, furniture purchased is selected according to the image the firm desires to express. Accordingly, furniture that is used in the customer service area differs from that used by the staff or management. These multinationals strive to impress their customer with an esthetical figure of rapid service, spruceness, modernism and professionalism. Furniture used in the "A" segment is designed using the colors and textures that distinguish each particular company, even though materials may remain the same, particularly the amalgamation of wood and metal. Many large overseas firms with subsidies or call centers in India have a corporate purchasing policy for furniture and the buying decision is made centrally at their head offices rather than locally. These corporations often call for standardized offices worldwide right down to the type of carpet used. This creates challenges for the Indian furniture manufacturers in contending with existing suppliers. Users in category "B" settle on leverages of office furniture principally based on price. Consumers in this segment tend to be especially price conscious due to generally low per capita incomes, a frugal mindset, a high propensity to save and buying primarily for need-based consumption. Firms in this market segment do not resort to the services of an architect and attempt to bargain the bare essentials, without much consideration for esthetics.

The decision-making process in furniture buying is a convoluted process that necessitates the consideration of constraints such as budget, available space, time investment, disposal of currently usable items, match with existing furniture and lifestyle [HUN 07]. Because of the significant expenses and lengthy product lifecycle of furniture, users have to make tough trade-off decisions about crucial factors such as price, style, quality and functionality. Moreover, after the choice is made, consumers often feel uncertain about whether they have made the correct choice.

4.4.2. *Distribution channels*

There is an excessive consumption of office furniture in the western and the northern markets of India, namely Mumbai and New Delhi, the latter being the leader. As represented in Figure 4.6 [IND 07c], Indian manufacturers employ a three-tier selling and distribution structure, that is, distributors, wholesalers and retailers. The gross profit margin for a retailer is 10–15%, followed by distributor (4–5%) and wholesaler (3–4%).

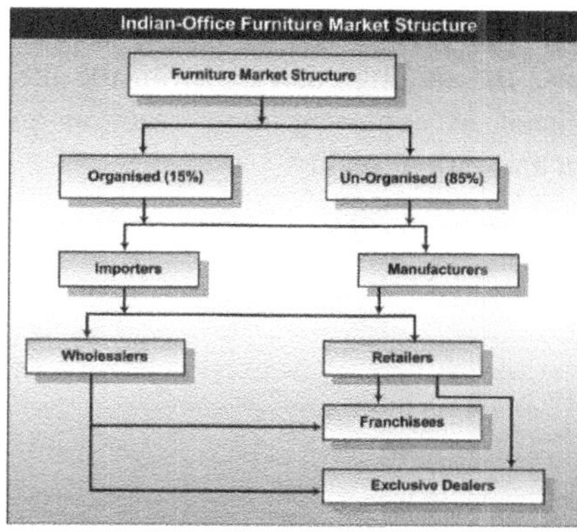

Figure 4.6. *Distribution chain for office furniture industry*

Indian companies sell furniture through distributors as well as directly to consumers. The leading office furniture importers also manufacture and sell their products via their own commercial offices and showrooms in all the large cities in India. Globalization has seen that imported furniture has made serious encroachments into the Indian markets, which has been a wakeup call for the office furniture manufacturing industry [IND 07c]. Figure 4.7 represents the schematic diagram of the value chain of Indian office furniture industry highlighting some of the internal and external factors associated with it.

4.4.3. *Data collection*

On the basis of the literature review, a questionnaire including both close- and open-ended questions was designed. The questionnaire has been divided into two main parts. The first part focuses on the demographics of the selected samples for the research, consumers' and stake holders' motivation for implementation of E-PD by employing online customization, whereas the second part concerns their opinions on implementation E-PD by employing online customization and virtual reality along with the evaluations of the benefits and pitfalls of the same. The questionnaire comprised 90% of close-ended enquiries. Those questions embraced (see section 4.9) five-point interval rating scale questions, which were ranked from "strongly disagree", "disagree", "neutral", "agree" to "strongly agree". In addition, semi-structured interviews are scheduled as informal interviews with the architects and dealers of the small office furniture businesses. Prior to semi-structured interviews, four pilot interviews were conducted to examine the research approach and the findings derived from the literature review. Pilot interviews also assisted in exploring their reaction and opinions toward online customization and virtual reality.

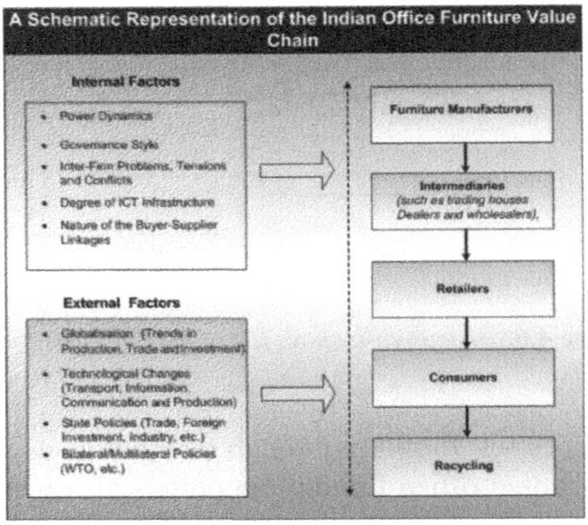

Figure 4.7. *Indian office furniture value chain*

The firm's selection aimed at the smaller dealers/architects who are successfully operating in the Indian office furniture market. Eight different small firms and three independent architects were selected as to be appropriate for the study, since they were not reluctant to share the postulated information. The project attempts to employ the firm's valuable experience in order to explore the viability of online customization and virtual reality in E-PD to improve customer satisfaction in the office furniture segment.

The interviewees were selected among the professionals: (1) managers/architects who do the business development for the firms, and (2) designers and the IT professionals who work in the office furniture industry. The motivation for reaching these professionals was to tap into the knowledge and "privileged information" that they share as a result of their extended years of experience in advising and working with a range of small office furniture manufactures and consumers.

4.5. Data analysis

This research project was conducted based on the rules of thumb furnished by Myers [MYE 97] via the following four stages:

1) *Ascertaining the current situation*: this was accomplished by the analysis of office furniture industry, using Political, Economic, Social, Technological, Legal and Environmental (PESTLE), Strengths, Weaknesses, Opportunities and Threats (SWOT) and Porters forces analysis.

2) *Accumulating information about the context of the present situation*: this was realized by information collated by interviews and citing documents and other sources available for the industry.

3) *Assemblage of specific data*: critical review of the success factors and consumer expectation for the office furniture firms and Indian market, by compiling and collating more reports and documents, along with the survey questionnaire.

4) *Exhibiting an analysis of findings and recommendations*: an exhaustive analysis is executed by employing the findings from the surveys, interviews and various frameworks from literature review. A modified conceptual model is proposed for achieving better customer satisfaction by employing new E-PD tools – online customization and virtual reality. SWOT and value chain analysis of the proposed conceptual model is realized equating the traditional model.

4.5.1. *Response of questionnaires*

A total of 57 individuals were offered the questionnaires via email. Of them, a total of 21 questionnaires were returned, that is a 37% response rate. However, 5 of the responses were incomplete and hence discarded from the survey and the remaining 16 responses were taken into consideration while compiling the analysis. The main reason for the lower response rate was the fact that all the questionnaires were sent through emails and so some people perceived it to be a spam mail. In general, most of the respondents completed sections A1.1 and A1.2 of the questionnaires.

4.5.2. *Findings on section A1.1 (demographics): getting to know the respondents*

A breakdown of the respondents' firms is presented in Figure 4.8. Majority of them are small proprietary establishments. Twenty five percent of the respondents are the medium-sized businesses. In terms of percentage, they along with the small firms make up to 69% of respondents.

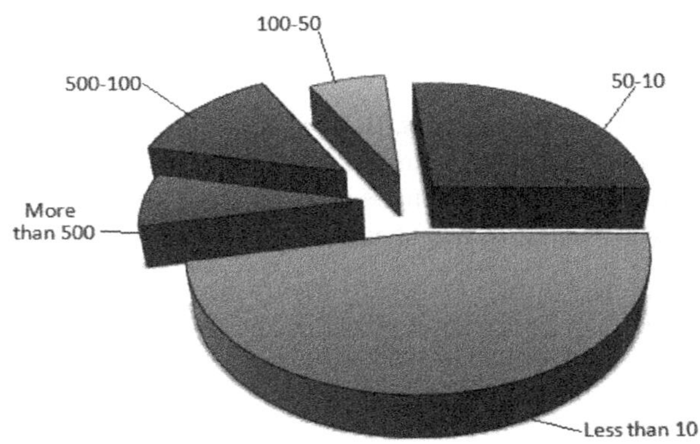

Figure 4.8. *Size of the businesses*

As shown in Figure 4.9, in terms of their position in the company, the majority of the respondents, about 31%, are designers and 25% of the respondents are marketing/sales executives having IT exposure. On the contrary to the interview respondents where three of the five respondents are proprietors, only one proprietor responded with the questionnaire.

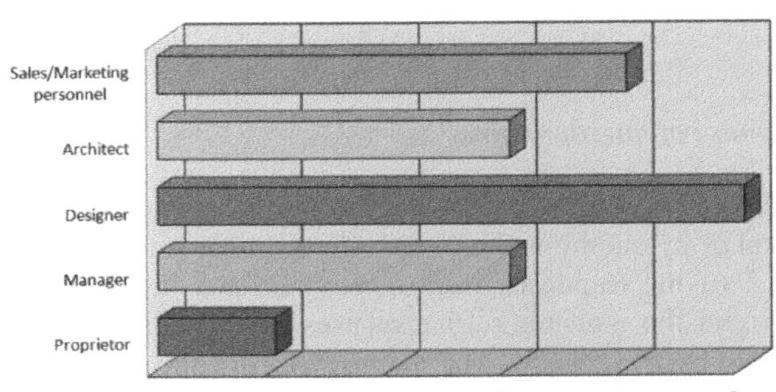

Figure 4.9. *Details for the respondents' positions*

More than 70% of the respondents have been working in the office furniture sector for more than 5 years. As shown in Figure 4.10, those with less than 24 months of experience make up only 12% of the respondents.

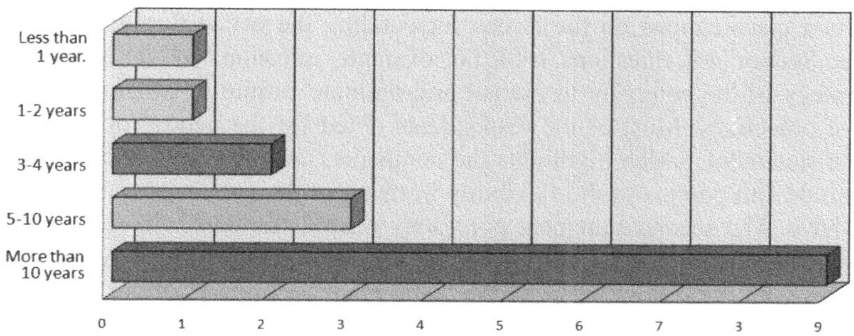

Figure 4.10. *Details for experience in office furniture market segment*

As illustrated in Figure 4.11, most of the respondents indicated that they have communications with the customers at least once every 2 weeks, thus they are very well aware of the market. In terms of breakdown, 38% of the respondents interact with consumers on a daily basis while another 31% have a close contact with customers between two to three times per week. This depicts that these responders are the correct target for replying to the questionnaires.

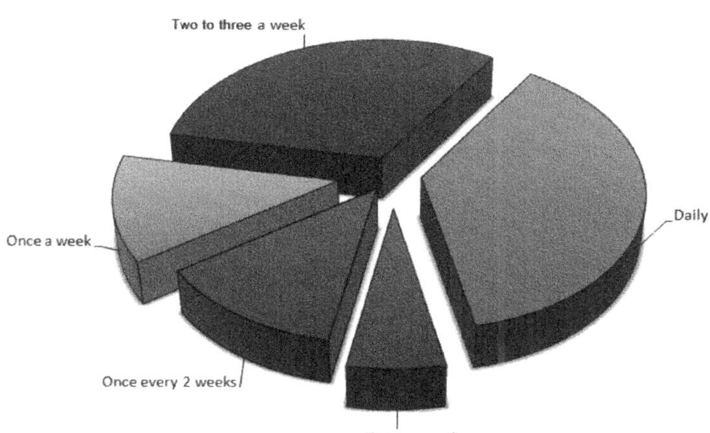

Figure 4.11. *Details for frequency of customer contact*

4.5.3. *Findings on section A1.2 (opinion): how they feel about these statements*

Figure 4.12 illustrates the responses to the statements that were drafted to obtain clarifications on the issues surrounding the office furniture business (see Section 4.9, questions 1–5). For example, question 1 (Q-1) is about the strategy of the company to market and promote furniture, which meets unit cost objectives. Most of the respondents opted for the neutral option in the first statement, which highlights the companies' policies toward cost-cutting attitude and points out the flexibility in terms of the policies that they intend to have. The second statement got a very neutral response, although 19% of responses are indicating "strongly disagree", overall it clearly suggests the issues with the excess furniture are being made redundant. About 58% (38% agree and 19% strongly agree) of the responders designate that they firmly believe that customers are becoming specifiers as they do not accept that the furniture they require cannot be made somewhere else. Statement 4 having 38% neutral responses, 25% agreeing and 13% disagreeing, respectively, again explicitly remonstrates the diplomatic and compromising policies adopted by most of the office furniture firms. It is interesting to note the 44% of responses strongly disagreeing and 0% strongly agreeing to the statement 5 that inquires about the current position of the firms to furnish the means for the customer to design their own products. This undoubtedly shows a wide gap that exists in the current situation.

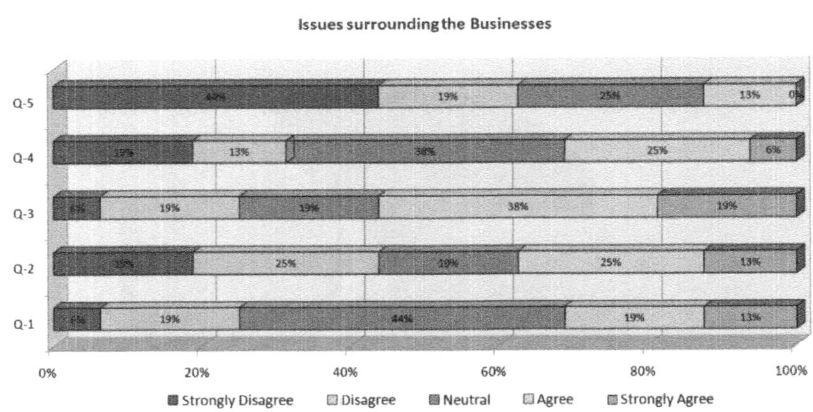

Figure 4.12. *Issues surrounding the business (responses to questions 1–5)*

According to the responses to business policies (question 6–8) shown in Figure 4.13, only 19% of the total responses disagree with statement 6, which indicates that most of the firms already know the benefits offered by the mass-customization toward the customer satisfaction while guarding the business interests in this competitive environment. Yet, again 69% of the respondents for statement 7 clearly elucidate the requirement for customization in office furniture design. However, respondents have different views on whether the customers should be allowed to actively participate in the designing of the furniture, as clearly evident from the responses of statement 8 in which 21% responders disagreeing and 25% communicating the neutral view on the approach. This section of the survey again highlights an opportunity for the office furniture firms to induce the customization approach in designing the furniture to improve customer relations.

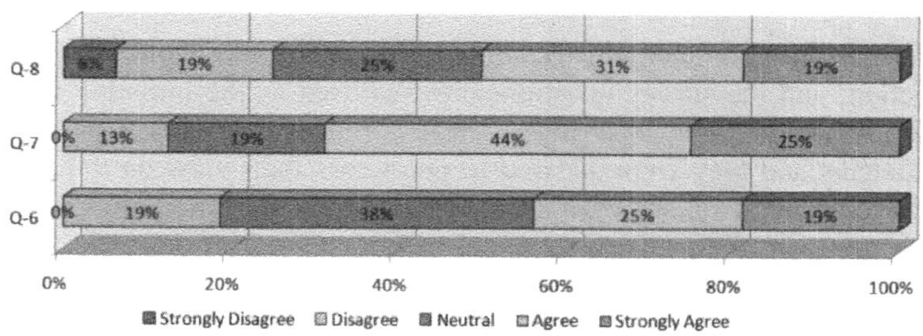

Figure 4.13. *Business policies (responses to question 6–8)*

As represented in Figure 4.14, the majority of respondents (70%) responded positively to statement 9 saying that the Internet could be employed as a handy tool for providing customers with an economical designing portal. In response of statement 10 regarding the utilization of IT and e-business solutions, 19% of the respondents inferred the negative replies coupled with 25% neutral responses; one of the reasons for these responses can be the lack of awareness of the respondents toward the core benefits of the e-business. Approximately 52% of the respondents agree on the benefits offered by the E-PD in terms of their overall business strategy, as clearly indicated by the responses collected for statement 11. More than half of the respondents indicated that they agree with statement 12, which points toward the value added by virtual reality toward consumer behavior while making the purchasing decision. Forty-four percent of the neutral responses toward statement 13 elucidate some concerns relating to the amalgamation of virtual reality and online customization in the E-PD of office furniture.

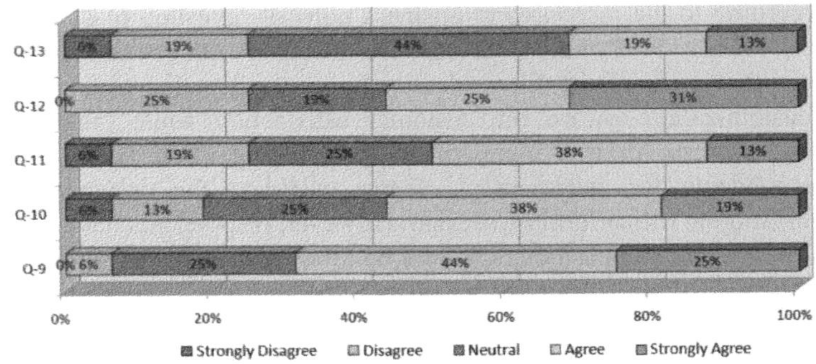

Figure 4.14. *Feasibility of online customization and virtual reality (responses to questions 9–13)*

As already mentioned, the triangulation approach was applied for the data collection, and the survey questionnaire was used as the initiating template for the semi-structured interviews. These pertinent questions were given to respondents and they were allowed to put across their view on the research objective. A total of six individuals took part in the interviews, typically lasting between 15 and 20 min over a 2-month period from June to July 2008. Data received from these semi-structured interviews yielded fresh insights into a wide range of issues. Overall, most of the interview results are in accordance with those found in the survey questionnaires and support the survey data analysis and results.

4.6. Discussions and further analysis using PESTLE

As shown in Figure 4.4 (the research design), this section deals with the discussion and analysis of the findings. Based on the market research, the PESTLE analysis of industry will be carried out, and the core issues regarding the SMEs furniture firms in India and the critical factors are finally determined by analyzing the gaps.

4.6.1. *Political and legal factors*

The political factors impacting the office furniture industry of India are shown in Table 4.1. The Indian office furniture sector is moderately influenced by the policies adopted by the government and the big firms especially with respect to import polices.

Political/legal factors
– According to recent surveys of American corporate executives, India is conceived to be the second best country after China for the business investment. India's "growing economy/business opportunities" and "low overall/operating costs" are attracting the worldwide capitalists to venture to India for business expansion [WAL 08].
– Some of the larger computer and office technology manufacturing sites in India are in joint venture with foreign corporations. Several factors have aided attracting and retaining these foreign investors. One prime aspect is the boost imparted by local, regional/national inducements.
– With the globalization of the Indian economy and dependent upon comfortable balance of payments status, the Government of India has liberalized the Import Policy and virtually all checks on imports have been lifted [IND 07c].

Table 4.1. *Political and legal factors affecting Indian office furniture sector*

4.6.2. *Economic factors*

The economic factors impacting the office furniture industry of India are listed in Table 4.2. In general, the office furniture industry is closely bound with the business cycle (i.e. ups and downs and financial performance) of its customers. Economic success or downturn in the industry is contingent upon the position of its customers' overall fiscal performance. When "times are good" and businesses grow, the acquisition of furniture, interiors and new office spaces becomes more important and vice versa.

Economic factors
– The growth in the furniture sector is closely linked to the growth in the economy. Today, when the Indian economy is growing at over 8%, the office furniture is growing at double-digit rates. The construction industry is flourishing as the government invests in infrastructure improvement and expansion; these ongoing expansions will open up a plethora of opportunities for companies in the furniture goods.
– Rising raw material costs and high oil prices have altered the distribution costs for suppliers/distributors of the office equipment.
– According to Knight Frank India research, the upsurge in demand from the IT/ITES and BPO sector has led to roughly 20–40% growth in capital values for office space in the last 12–15 months across major metros. This evidently affects the market for office equipment.
– The number of jobs in the private sector is escalating following the economic boom. This has led to transformations in the design of office equipment, the adjustability of furniture and the layout of offices.

Table 4.2. *Economic factors influencing Indian office furniture sector*

4.6.3. *Social factors*

One common social issue, confronted by many of the smaller office furniture manufacturers, is focused on the attraction and retention of employees in a competitive labor market. The social factors affecting the office furniture industry of India are presented in Table 4.3.

Social factors
– Attracting people with the right skill sets and retaining them is the key to success. SMEs find it hard to attract reasonably priced, well-trained executive teams, and other workforce, when there are strong stresses on prices due to the competition from firms outside of India where remunerations and benefits may be very low. These businesses are required to be able to present good salaries and welfares to draw the class of employees they desire. Another ingredient that increases this complexity of pulling staff is the large number of work opportunities available for these trained workers outside the office furniture sector [DAT 08].
– Ergonomic design of office furniture is also becoming mandatory. To reduce the incidence of repetitive strain injuries, to improve employee health and to scale down workplace stress, office equipment should be designed bearing in mind the Indian ergonomics standards.

Table 4.3. *Social factors influencing Indian office furniture sector*

4.6.4. *Technological factors*

The miniaturization trend of electronic and IT equipment has set aside office equipment to become more compact and lightweight, and has changed the magnitude of the range of portable products. Computers, laptops and display screens are becoming more compact, reducing desk space requirements. Some other technological factors affecting office furniture sector are shown in Table 4.4.

4.6.5. *Environmental factors*

Environmental issues are turning more extreme. Public sector organizations are expected to exhibit that they are environmentally aware, and so are prominent private sector companies. This can affect buying decisions in the office equipment market, and manufacturers are investing heftily in amending their environmental performance. Some of these environmental factors are illustrated in Table 4.5.

Technological factors
– The dramatic growth of the Internet is transmitting to the Indian market. The buyer nowadays gets an inexhaustible and predominately unrestricted stream of product information from diverse sources. The precipitous volume of media exposure shapes the very conception of communication. Consumers are well informed about products in new and very creative ways from spam email to glimpses of brands in movies.
– Excessive media influence has taught consumers that they have alternatives and can acquire what they want no matter. If the required product is not available on the shelf of the local retailer, they go online to the Websites or surf for imported products.
– With the ever-growing influx of new technology in the office, that is personal computers, laptops, printers and rechargers for the mobile devices, there is a rising demand from the office furniture industry to respond by rendering integrated products to manage all the power and data requirements besides exploring new furniture opportunities to support mobile and wireless technology.
– The recent growth of fast network accessibility and Web 3D technologies has prompted virtual reality-based tools as a medium for conveying product information and exploration with substantial values to office furniture customers.
– Presently several internal business process management systems, such as enterprise resource planning (ERP) tools, are being developed. With the support of appropriate IT and e-business solutions, organizations can realize extensive cost reductions and time saving, by integrating all aspects of the supply chain [FUR 08].

Table 4.4. *Technological factors influencing Indian office furniture sector*

Environmental factors
– Institutional and government clients often require that office furniture products constitute a certain minimal percentage of reprocessed materials in their manufacture.
– It is evident from the past trends that marketplace expectations in India are continuously evolving. New standards or regulations compliance may emerge in particular associated furniture products and present a possible business risk. These switches in market expectations could pose future challenges or material costs, process or technology changes and certification costs.
– There are increasing pressures from the government to design, build and manage constructions and operations in an environmentally sound manner. This motion has rendered market pressures on furniture manufacturers to furnish products that can assist architects, interior designers and space planners to meet these demands. The volume of so-called "green buildings" in India is soaring year by year with the global pressure of reducing emissions [IND 08b].
– Recyclability of material is another primary issue for the manufacturers and consumers in the industry. It primarily concerns the possible volumes of waste created by office furniture goods at the end of their operational life [FIV 03].

Table 4.5. *Environmental factors influencing Indian office furniture sector*

4.6.6. *Critical factors for decision-making over E-PD implementation*

Four groups of factors including internal and external factors, specific factors and desired characteristics, most of them being qualitative and difficult to quantify, could influence a decision over entry of a given product/target country [HOL 07]. By adapting these influencing factors along with the research findings, the positive/negative impacts of various market factors on the decision over adoption of the E-PD for the office furniture segment in India can be found as illustrated in Figure 4.15. Most internal/external factor and desire characteristics, including control of resources and flexibility, have positive impacts on the E-PD implementation. However, all of the specific factors including opportunity cost, the tacit nature of knowledge and financial capabilities negatively influence the E-PD implementation. Similarly, furniture capability has a positive effect whereas the product differentiation advantage has a negative impact on the E-PD implementation.

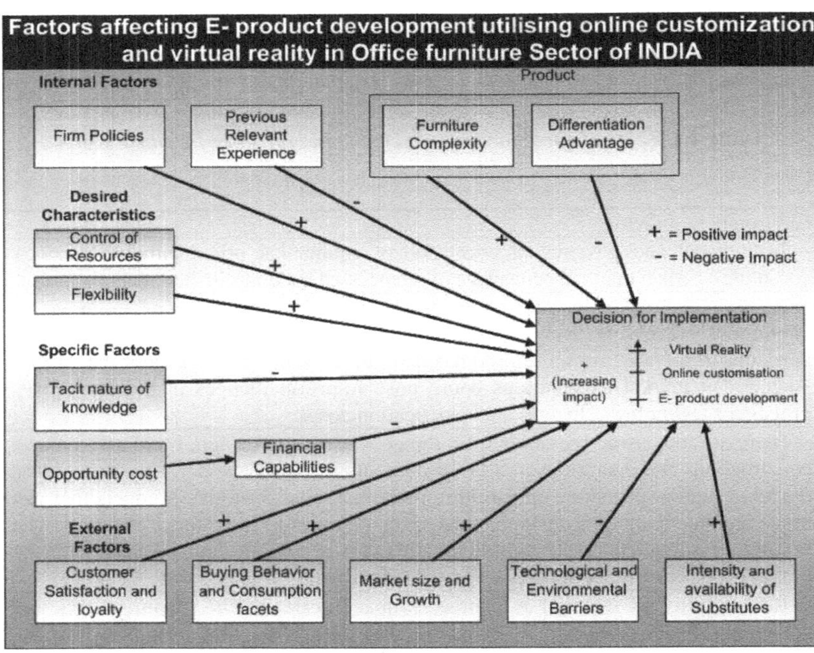

Figure 4.15. *Factors affecting E-PD utilizing online customization and virtual reality*

4.7. Conclusions

Customization along with virtual reality technologies enable the users to have dynamic interaction with objects. The main issue in office furniture leverages is the inability of customers to view their self-designed products in their own premises. Demonstrations at brick-and-mortar stores do not resolve this crisis as consumers still cannot match options with their currently owned site décor. However, close inspection of individual pieces of furniture with a personalized environment can be obtained when the furniture prototypes are rendered in 3D real-time computer graphics. Thus, customization and virtual reality are viable tools that offer possible resolutions to many identified problems facing the current office furniture industry of India. The research outcome foresees the future of the office furniture industry as follows: consumers go online, try novel/innovative furniture items, analyze the esthetics of their own site with new furniture items before making a decision. The furniture selection made by customers will be installed at customer's premises after physical testing.

Realizing the gains that online customization and virtual reality assures is not going to be an easy task, as there are many formidable obstacles to overcome, for instance an entrenched attitude that revolves around the diminution of labor and input costs as a competitive advantage rather than following a knowledge intensive trajectory. Furthermore, closed-minded administration of some of the interview respondents highlighted that the main barrier, for utilizing Internet as a powerful tool, is the lack of understanding about how their products may be designed online or endorsed by Internet technologies. In most cases, firms lack technical knowledge, and the cost of participation in E-PD is too high for them. While most of the respondents desire to use online customization and virtual reality as tools, they want these to be intuitive for the user, rather than highly complex and necessitating specialist salespeople to administer these products. The indispensable training, for these so-called specialists, to generate the expertise for firms leads to exorbitant costs and a long-time investment, both of which affect heavily the limited resources accessible to these companies. Customizing and virtual prototyping can alter the office furniture E-PD process in ways that economize costs, cut time to market and step-up receptiveness to consumers' taste changes.

In conclusion, adoption of E-PD and virtual reality tools embodies a viable option for small office furniture selling firms seeking to augment their customer base, boost customer satisfaction and place themselves alongside big firms and small niche players. However, the costs of enforcing these technologies are a large prohibitive factor. The majority of these enterprises strive to subjugate expenditures on software management. Owing to the complexity and simultaneous benefits of these systems, small firms must adopt long-term visions in terms of return on investment and abandon their short-term, cost-cutting measures in these critical areas where it has been demonstrated that the investment would be fruitful.

In particular, the research findings should address the key policy-makers and representatives of the Indian office furniture, especially firm managers, decision-makers in product development and marketing, IT and e-solutions managers.

Even though the process of compiling data was productive, the research work does have its own limitations in applying to the practical business world. It was realized that the office furniture market of India is vast and complicated. Within the limited time, inaccessibility to some crucial market data, besides lack of information directly from the consumers with regard to visual reality and customization could skew the outcomes, thus the opinions may not be comprehensive:

1) The data accumulated were not fully comprehensive and consequently some premises had to be made, based on the authors' own experience of the business, due to time constraints.

2) The current sample size for this study is quite small. Larger sample size of questionnaires and conducting more interviews could have improved the validity of the research. In a few cases of the questionnaire, the responses were vague and incomplete. Follow-up questions via telephone would have been the best way to resolve the issue.

4.8. Bibliography

[ABD 03] ABDI M.R., LABIB A.W., "A design strategy for reconfigurable manufacturing systems (RMSs) using the analytical hierarchical process (AHP): a case study", *International Journal of Production Research*, vol. 41, no. 10, pp. 2273–2299, 2003.

[ABD 04] ABDI M.R., LABIB A.W., "Grouping and selecting products: the design key of reconfigurable manufacturing systems (RMSs)", *International Journal of Production Research*, vol. 42, no. 3, pp. 521–546, 2004.

[AND 90] ANDERSON D.M., *Design for Manufacturability, Optimizing Cost, Quality and Time-to-Market*, CIM Press, Lafayett, CA, 1990.

[ANS 03] ANSARI A., MELA C.F., "E-customization", *Journal of Marketing Research*, vol. 40, pp. 131–145, 2003.

[AVL 00] AVLONITIS G.J., KARAYANNI D.A., "The impact of internet use on business-to-business marketing examples from American and European companies", *Industrial Marketing Management*, vol. 29, no. 5, pp. 441–459, 2000.

[BAR 03] BARDAKCI A., WHITELOCK J., "Mass-customisation in marketing: the consumer perspective", *Journal of Consumer Marketing*, vol. 20, no. 5, pp. 463–479, 2003.

[BEL 09] BELAY A.M., "Design for manufacturability and concurrent engineering for product development", *Engineering and Technology*, vol. 49, pp. 240–246, 2009.

[COV 07] COVA B., KOZINETS R.V., SHANKAR A., *Consumer Tribes*, 1st ed., Butterworth-Heinemann/Elsevier, Oxford, UK, 2007.

[DAT 08] DATAMONITOR, Marketline, India Country Profile, 2008.

[DUR 00] DURAY R., WARD P.T., MILLI G.W., "Approaches to mass customization: configurations and empirical validation", *Journal of Operations Management*, vol. 18, pp. 605–625, 2000.

[FIV 03] FIVE WINDS INTERNATIONAL, Sustainable Development Issues Scan: Office Furniture Industry, 2003.

[FUL 07] FULLER J., MATZLER K., "Virtual product experience and customer participation – a chance for customer-centred, really new products", *Technovation*, vol. 27, pp. 378–387, 2007.

[FUR 08] Furniture – Interim study report, January 2008. http://www.ebusiness-watch.org/studies/sectors/furniture/furniture.htm (accessed July 24, 2008).

[GOL 04] GOLDSMITH R.E., FREIDEN J.B., "Have it your way: consumer attitudes toward personalized marketing", *Marketing Intelligence & Planning*, vol. 22, no. 2, pp. 228–239, 2004.

[HEL 02] HELANDER M.G., JIAO J., "Research on e-product development (ePD) for mass customization", *Technovation*, vol. 22, no.11, pp. 717–724, 2002.

[HIG 01] HIGGINS A., "Designing with the new Internet", *Machine Design*, vol. 73, no. 14, pp. 90–93, 2001.

[HOE 06] HOEFFLER S., ARIELY D., WEST P., "Path dependent preferences: the role of early experience and biased search in preference development", *Organizational Behavior and Human Decision Processes*, vol. 101, no. 2, pp. 215–229, 2006.

[HOL 07] HOLLENSEN S., *Global Marketing*, 4th ed., Prentice Hall, Essex, England, 2007.

[HUN 07] HUNTER S.L., LI G., "Market competition forces: a study of the Chinese case goods furniture industry", *Forest Products Journal*, vol. 57, no. 11, pp. 21–26, 2007.

[HUR 00] HURWICZ M., It's ready when you are, 2000. Available at http://www.webdevelopersjournal.com/articles/virtual_reality.html (accessed 26 June 2008).

[IND 07a] INDIA & THE FURNITURE MARKET, Universal expositions, 2007. Available at www.ueindia.com/papers/media_aboutindia.pdf (accessed 26 July 2008).

[IND 07b] INDIA SECTOR PRESENTATION: FURNITURE, India brand equity foundation, 2007. Available at http://www.ibef.org/artdisplay.aspx?cat_id=259&art_id=18127 (accessed 10 July 2008).

[IND 07c] INDO-ITALIAN CHAMBER COMMERCE AND INDUSTRY, Overview of furniture sector in India, 2007. Available at http://www.perfspot.com/ docs/doc.asp?id= 27880 (accessed 30 March 2013).

[IND 08] INDO-ITALIAN CHAMBER COMMERCE AND INDUSTRY, India: export opportunities in a growth market, 2008. Available at http://www.ud.camcom.it/Risultati-ricerca/P46O0/india (accessed 30 March 2013).

[IND 08b] INDIAN GREEN BUILDING COUNCIL, "Green Building", *Indo-German Symposium on Energy Efficiency*, New Delhi, 2008.

[IND 11] INDO-ITALIAN CHAMBER OF COMMERCE AND INDUSTRY (IICCI), 2011. Available at http://www.ud.camcom.it/Risultati-ricerca/P46O0/india (accessed 30, March 2013).

[IYE 03] IYER G., PAZGAL A., "Internet shopping agents: virtual co-location and competition", *Marketing Science*, vol. 22, no. 1, pp. 85–106, 2003.

[JIA 06] JIANXIN J., HELANDER M.G., "Development of an electronic configure-to-order platform for customized product development", *Computers in Industry*, vol. 57, no. 3, pp. 231–244, 2006.

[KAW 11] KAWAKAMI T., DURMUŞOĞLU S.S., BARCZAK G., "Factors influencing information technology usage for new product development: the case of Japanese companies", *Journal of Product Innovation Management*, pp. 1–15, 2011. Available at http://onlinelibrary.wiley.com/doi/10.1111/j.1540-5885. 2011. 00848.x/pdf (accessed 30 August 2011).

[KOT 09] KOTLER P., KELLER K.L., *Marketing Management*, 12th ed., Pearson Prentice Hall, Upper Saddle River, NJ, 2009.

[LUR 07] LURIE N., MASON C.H., "Visual representation: implications for decision making", *Journal of Marketing*, vol. 71, pp. 160–177, 2007.

[MEN 99] MENDOZA M., "Collaboration beyond engineering", *Computer-Aided Engineering*, vol. 18, no. 7, pp. 24–27, 1999.

[MIR 07] MIRANDA M.J., KÓNYA L., "Customisation – moving customers away from the dull conformity of brand loyalty", *Managing Service Quality*, vol. 17, no. 4, pp. 449–466, 2007.

[MYE 97] MYERS M.D., "Qualitative research in information systems", *MIS Quarterly*, vol. 21, no. 2, pp. 241–43, 1997.

[OZE 03] OZER M., "Process implications of the use of the Internet in new product development: a conceptual analysis", *Industrial Marketing Management*, vol. 32, pp. 517–530, 2003.

[SAM 97] SAMLI A.C., WILLS J.R., HERBIG P., "The information superhighway goes international: implications for industrial sales transactions", *Industrial Marketing Management*, vol. 26, no. 1, pp. 51–58, 1997.

[SKJ 01] SKJELSTAD L., HAGEN I., ALFNES E., "Guidelines for achieving a proper mass customization system", *EurOMA International Conference on Operations and Global Competitiveness, Session: Production and Manufacturing Management*, SINTEF Technology and Society, pp. 1565–1572, 2001.

[SON 07] SONG K., FIORE A.M., PARK J., "Telepresence and fantasy in online apparel shopping experience", *Journal of Fashion Marketing and Management*, vol. 11, no. 4, pp. 553–570, 2007.

[STA 03] STADTLER H., KILGER C., *Supply Chain Management and Advanced Planning Concepts, Models, and Case Studies*, 4th ed., Springer, New York, 2003.

[STU 97] STUMP R.L., SRIRAM V., "Employing information technology in purchasing: Buyer-supplier relationships and size of the supplier base", *Industrial Marketing Management*, vol. 26, no.2, pp. 127–136, 1997.

[SYA 08] SYAM N., KRISHNAMURTHY P., HESS J.D., "That's what I thought I wanted? Miswanting and regret for a standard good in a mass-customized world", *Marketing Science*, vol. 27, no. 3, pp. 379–397, 2008.

[TSE 98] TSENG M.M., JIAO J., SU C.J. "Virtual prototyping for customized product development", *Integrated Manufacturing System*, vol. 9, no. 6, pp. 334–343, 1998.

[VIA 04] VIARDOT E., *Successful Marketing Strategy for High-Tech Firms,* 3rd ed., Artech House, London, 2004.

[WAL 08] WALL STREET JOURNAL, August 2008, India second best country for business investment: Survey.http://www.livemint.com/2008/08/01113603/India-second-best-country-for.html (accessed Aug 2008).

4.9. Appendix

4.9.1. *Questionnaire*

A1.1.

Please mark (√) against the appropriate box.

1) What is the size (in terms of number of persons employed) of your company?

i) More than 500

ii) 500–100

iii) 100–50

iv) 50–10

v) Less than 10

2) Which of the following best describes your designation?

i) Proprietor

ii) Manager

iii) Designer

iv) Architect

v) Sales/Marketing personnel

3) For how long have you been worked in the office furniture or relevant sector?

i) More than 10 years

ii) 5–10 years

iii) 3–4 years

iv) 1–2 years

v) Less than 1 year.

4) How often do you have to deal with customers directly?

i) Once a month

ii) Once every 2 weeks

iii) Once a week

iv) Two to three times a week

v) Daily

A1.2.

Please mark (√) against the appropriate box.

Rating scale				
Strongly disagree	Disagree	Neither agree nor disagree	Agree	Strongly agree
SD	D	N	A	SA

S. No.	Statements	SD	D	N	A	SA
1	The strategy of your company is to market and promote furniture that meets unit cost objectives.					
2	Firm has some styles of furniture that customers have not accepted, resulted in a costs of holding immense inventories.					
3	Customers are aware of what they want and they would not accept that it is not available. If they do not find what they intent, they would go to Internet and try to search it abroad.					
4	The policy of your firm is to offer furniture that is tailored and shaped to fit customers' demands.					
5	Presently your firm furnishes a platform and tools and "rents" out to customers the means to design their own furniture					
6	Employing mass-customization in the office furniture market could extend service and could help achieving a greater level of customer satisfaction.					
7	The firm should launch furniture aimed at customization for developing customer loyalty.					
8	Involving customers into designing their own furniture could enable firm to respond quickly to customer needs.					
9	Internet can be proved as an economically feasible device for the customizable 3D modeling, which can enhance the efficiency in the collaborative furniture designing.					

10	With the support of appropriate IT and e-business solutions, organizations can realize extensive cost reductions and time savings, by integrating all aspects of the supply chain.					
11	E-product development could assist in interpreting customers' preferences and decision-making behavior, which can lead to better product development, marketing and distribution management for the furniture industry.					
12	Virtual reality posses peculiar features that grant users to analyze products closely and allows in quick decision-making, all of which are valuable for customer satisfaction.					
13	Internet coupled with virtual reality tools can speed up the testing and validation while diminish the cost of same for new customized products.					
14	Your firm has constraints with respect to cost time and skill for developing infrastructure for the online customization and virtual reality.					
15	To offer consumers customized furniture at nominal prices, mass-customization practices such as reducing costs through lowered production costs, bringing down R&D costs and diluting inventories are a must.					

Chapter 5

Struggling for Survival and Success: Can Brazil's Defense Industry Help Foster Innovation?

Innovation is one of the main processes by which organizations can create and add value to their business. It also plays a critical role in the health and prosperity of nations. Nevertheless, innovating is not easy because it depends on various factors that can create a favorable environment or undermine its occurrence.

The objective of this chapter is to address the mechanism of offset agreements, which are widely used in the acquisition of defense products, as a means to prevent some major barriers to innovation in the Brazilian Defense Industry (BDI).

This chapter begins with a brief literature review of the main topics concerning technological innovation, the main barriers to innovation and the role of offset contracts in the innovation process. The results obtained from empirical research that is carried out in the context of the BDI are then presented and discussed.

Chapter written by Alex Lôbo CARLOS and Regina Maria DE OLIVEIRA LEITE.

5.1. Introduction

The study of technological innovation had its origins in the industrial, commercial and private sectors: first, in the area of production and subsequently in services. It was intensified after the publication of the first edition of the "Frascati Manual" [ORG 63] that introduced standard practices for surveys on Research and Development (R&D). More recently, innovation studies have extended to other contexts, such as the public sphere and organizations. Governments all over the world are adopting innovation strategies and tools in order to improve the performance and the effectiveness of public organizations as well as private companies.

One of the most important challenges that governments are facing today is the need to stimulate an innovation-friendly environment and to remove barriers to its implementation. By doing so, they will be able to improve a nation's economic growth and competitiveness. A good example of a country's innovative power can be found in the discourse of the American economist Amara [AMA 90]. In 1990, he estimated that 60–80% of the American GDP, *per capita*, came from a residual component of innovation and technology [AMA 90]. This means that innovation plays a crucial role in economic growth and is becoming even more important in the current context of globalization with an unprecedented level of competitiveness.

In Brazil, between 2000 and 2010, the innovation efforts and investment in R&D came mostly from the government, while in many developed countries private companies take the main responsibility [MIN 12, ORG 12b]. Regardless of who the principal investor is, the need for innovative strategies and solutions is imperative for countries and organizations. In today's crisis, it is urgent to maximize results and minimize governmental and organizational expending of resources.

With this in mind, some countries are currently using the offset agreements to reach their objectives. Although the practice of offset agreements is not new, in recent years we have witnessed a growing number of countries applying offsets and other compensation measures in major defense contracts [VER 09]. Governments are making use of their unique purchasing power to develop their domestic industry by capitalizing on offset benefits. The offset agreements can take different forms in order to help innovation, such as coproduction, credit assistance, investment, license production, purchase, subcontracting, technology transfer, training and others, which will be explained later.

5.2. Innovation as a driver for success and its common hurdles

The concept of technological innovation is central to this chapter because of its capacity to generate financial results and stimulate competitiveness. It is defined as the introduction to the market of a new product or a new service or a significant technological change to a product or to a process [ORG 05].

The existing body of literature around the issue of innovation clearly supports its relevance for organizational survival and success. In 1991, Nayak [NAY 91] highlighted the importance of product innovation and its contribution to profit improvement and organizational revenue. A decade later, Schacht [SCH 00] argued that innovations are a strong determinant of growth in small and medium enterprises. In 2009, Madrid-Guijarro *et al.* [MAD 09] said that innovation is a key factor in competitiveness of countries and companies. According to Sarkar [SAR 10], organizations have to develop new products and services in order to increase their financial results. Freire [FRE 00] argues that a company has to put together three elements to be innovative: opportunity, desire and capacity.

Bearing in mind the above-mentioned arguments and evidence, governments are taking some action to promote a favorable environment for innovation. Such government measures are intended to increase business efficiency, and consequently a nation's welfare. Above all, these public policies are meant to help support projects that would not have happened otherwise [LÖÖ 05, GUP 08].

Innovation barriers have attracted much attention from researchers and practitioners. The main barriers that are reported relate to cost, knowledge, market and institutional factors [ORG 05].

An empirical study carried out by the well-known computer company IBM has outlined the main obstacles and constraints facing innovation. They can be summarized as follows [IBM 06]: restrictions imposed by the government, economic instability, unsuitable technology, labor force problems, cultural restrictions, investment limitations, lack of expertise and restricted access to information.

Regardless of the barriers presented above, human capital plays a leading role in innovation. There seems to be a consensual agreement that human capital is more important than the innovative infrastructure or large innovation investments [JOB 11, CAR 11]. Human capital is central to innovation because inventions require employees with expertise and creativity, and there will be no innovation unless the market validates the inventions.

The level of human capital depends heavily on education and other related issues, such as training and development. Qualified people are a prerequisite for successful innovation, and both developed and developing countries are taking measures in order to increase the number of skills and qualifications of their workforce.

5.2.1. *A brief portrait of innovation in Brazil*

Brazil is currently the seventh largest economy in the world, but performance indicators in the sphere of education, health and social justice present a less favorable and quite upsetting picture.

According to the "Education for All Global Monitoring Report 2011" [UNE 11], Brazil ranked 88th among 127 countries.

A recent Organization for Economic Co-operation and Development (OECD) report shows that only 11% of the Brazilian population, between 25 and 64 years old, hold a university degree. On average, 28% of the other OECD members have achieved this level of education [ORG 12a]. According to the Human Development Index (HDI) and the United Nations Development Program (UNDP), Brazil ranked 84th in 2011 [UND 11].

When we look at the innovation data, the picture remains pretty much the same. Brazil's performance is relatively poor when compared with other countries. According to the "Global Innovation Index 2012" [DUT 12], Brazil occupies the 58th position (Table 5.1), far below expected, especially if we consider the country's dimension.

Nevertheless, this scenario can be reversed in the near future because the government has been made aware of R&D's enormous potential and is now investing in education, as well as in R&D infrastructure and projects. As a result, during the past years, the country has experienced a remarkable economic growth. The "Global Innovation 1000" [JAR 12] report shows that, in 2011, seven Brazilian enterprises ranked among the thousand most innovative enterprises of the world. It also highlights the efforts made in 2011 by these Brazilian enterprises to increase R&D investment. According to the "Global Innovation 1000" report, Brazilian companies have witnessed an increase of 54% while the 1,000 companies have increased 9.6%.

However, this does not accurately depict the reality of most Brazilian companies. In fact, a large majority of companies tend to perceive R&D investment as extremely expensive.

Country	Ranking position
Switzerland	1st
Sweden	2nd
Singapore	3rd
Finland	4th
The United Kingdom	5th
The Netherlands	6th
Denmark	7th
Hong Kong (China)	8th
Ireland	9th
The United States of America	10th
Luxembourg	11th
Canada	12th
New Zealand	13th
Norway	14th
Germany	15th
China	34th
Portugal	35th
Russian Federation	51th
South Africa	54th
Brazil	58th
India	64th

Table 5.1. *Global innovation index 2012 [DUT 12]*

In Brazil, R&D investment comes mostly from the government when compared to other countries. Long periods of high inflation rates and uncertainty might have contributed to the low investment of private enterprises. Surviving hard times was their first priority. To overcome the R&D investment shortage, the Brazilian government has taken a leadership role and started strengthening R&D policies to innovate and increase the country's international competitiveness.

In 2011, the law no. 12.545 of 14th December 2011 [BRA 11] has changed the name of the Ministry of Science and Technology to Ministry of Science, Technology and Innovation, thus highlighting the importance of, and giving more visibility to, the subject of innovation in Brazil. The program "Science without Frontiers" ("*Ciência sem Fronteiras*") was launched in the same year, and was intended to give 100,000 students the opportunity to pursue a higher-level degree at the top international universities or helping them to become researchers. The estimated total size of this investment is 1.5 billion dollars. This massive increase in educational and research expenditure is a necessary, although insufficient step toward innovation success. Brazil urgently needs to identify the barriers that prevent innovative and the success of innovative enterprises in order to effectively fight against them. Otherwise, such obstacles will destroy any innovation initiative [OWE 11].

The Brazilian Institute of Geography and Statistics (*Instituto Brasileiro de Geografia e Estatísitica* – IBGE) has promoted, since 2000, the so-called Innovation Research program (*Pesquisa de Inovação* – PINTEC) on a national level. Now in its fifth forthcoming edition, the program has already issued four reports (2000, 2003, 2005 and 2008) [IBG 00, IBG 03, IBG 05, IBG 08]. The next publication will cover the three-year period from the beginning of 2009 to the end of 2011 [IBG 12], and its data collection was started last June.

Those reports are based on the "Oslo Manual" [ORG 05] standard, and one of the main purposes is to evaluate the incidence of barriers to innovation from the enterprises' point of view.

The main barriers to innovation that have been reported by the participants are the lack of qualified people (internal nature), the shortage of funding sources, the excessive economic risks and high costs of innovation (economic nature). The above-mentioned barriers are in-line with those figures published in other reports, such as the OECD [ORG 05] or IBM [IBM 06].

A noteworthy aspect is that the various reports have consistently shown those barriers as the most important obstacles over the years. In spite of the recent actions, investment and efforts made by the Brazilian Government, there is still much work to be done to overcome such hurdles.

5.3. Offset agreements: concepts and applications

The five countries (Brazil, Russia, India, China and South Africa) known as BRICS have been working in order to reduce the barriers to innovation. Some of these countries are developing public policies taking into account their own specificity, as demonstrated by the "STI 2012" report [ORG 12b]. This OECD shows that despite the crisis period and consequent decline in innovation activities at the OECD area, emerging countries such as Brazil and India, and mainly China, are prioritizing innovation on their political agenda.

In this regard, governments have adopted policies that have already been used in other countries, such as the practice of offset contracts or agreements. For example, South Africa, Brazil and India have signed offset agreements in order to gain access to missing or incomplete competencies and develop the domestic sectors.

The legislative measures adopted by the BRICS are a clear demonstration of this awareness. In this regard, India revised its defense offset policies in 2011 [MIN 11], and Brazil, which was previously applying the *Portaria* 764/2002 of the Ministry of Defense [MIN 02], has published the law no. 12.598/2012 in 2012 [BRA 12] putting more emphasis on offset issues.

Some argue that the purchasing contracts that involve compensation usually include additional costs to buyers that may vary between 7% and 10% [BRA 04]. Despite the controversy surrounding the impact of offsets in reaching their declared goals, more and more countries all over the world are using their purchasing power for gaining access to knowledge and technologies that they otherwise could not afford.

The offset agreements are complex, and can be understood as part of a larger net of business practices called countertrade (see Figure 5.1).

Brauer and Dunne [BRA 04] describe the offset agreements as a growing practice within the international trade of weapons that requires the seller to compensate the purchaser, usually a government, through a reinvestment agreement, on behalf of the latter or some enterprises within that nation. The compensation value tends to be similar to that earned by the seller of the military equipment. The offset agreements commonly correspond to a percentage of the total commercial contract amount (in the Brazilian case, a minimum of 100% is required).

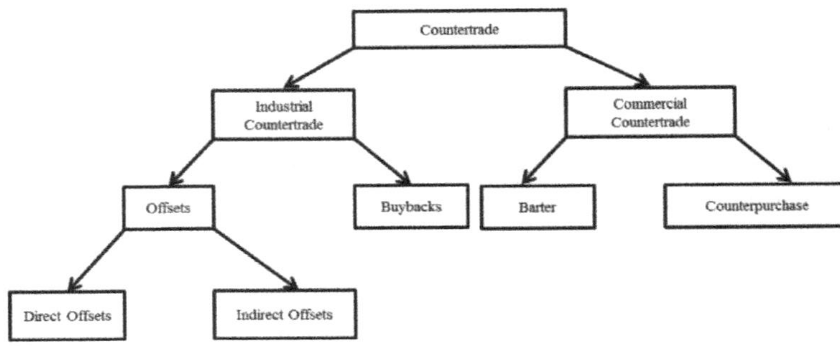

Figure 5.1. *Business practices (adapted from [BRN 98])*

The offset operations, which originated in the context of military acquisitions, demand the government's participation as one of the involved parties in such contracts. Offset is defined as "Industrialist compensation practices required as a condition of purchase in either government-to-government or commercial sales of defence articles and/or defence services" [BUR 12, p. 28]. Other synonymous words and expressions for offsets can be found in many dictionaries (e.g. Oxford, Cambridge and Collins): "something that counterbalances, counteracts, or compensates something" and "compensating, equivalent".

It is important to remember that without a commercial contract there is no offset agreement. An offset agreement is then a consequence of a commercial contract. Some parts are involved in the offset agreements. In general, we can find the seller (a company or country that sells the military material, item, equipment or system) and the buyer (the country or company located overseas that buys the military material, item, equipment or system). These concepts are related to the purchasing contract. When we talk about offset agreements, these parts turn into offset recipient or beneficiary (buyer) and offset offerer or debtor (seller).

5.3.1. *The early days of offsets*

After the Second World War, Americans invested significantly in the military reconstruction of impoverished Europe. Back then, offsets were widely used to help buyers reduce the expenses of the countries. Offsets have not only contributed to supplying Europe with military goods and expertise, but have also helped in producing jobs and transferring technology to the purchasing countries [NEU 85, JON 02, MOD 04].

The origin of offsets dates back to 1930, following the significant barriers imposed on international trade and the high external debt of countries after the global economic recession. To overcome the difficult situation, both governments and private enterprises started searching for recovery solutions that would help them emerge from the crisis. The volume of offset contracts has grown since its inception, and nowadays it is estimated that offsets account for around 20–30% of the international trade [HOW 10].

If the offset contracts are properly negotiated, they are expected to help countries reach their economic, social and geopolitical targets [IVO 04] that would otherwise take longer to be achieved.

Although these agreements have their roots in the context of purchasing military apparatns, such equipment has a technological component that is relevant to both military and civilian domains. The products and technology that simultaneously serve military and civil purposes are commonly known as dual-use technology [EUR 09]. For example, global positioning system (GPS), radar and the cell phone were originally conceived for military use, but today are part of everyone's daily routine. The potential dual-use of military products and technology is of great importance because more than 35% of the offsets negotiated globally refer to the transfer of technology [SUM 10].

5.3.2. *Types of offsets*

Offset contracts or agreements can commonly be classified as direct or indirect. In a direct offset, compensation is directly associated with the products or services being purchased [PAR 06], while in an indirect offset the benefits are directed toward activities beyond the scope of the purchasing contract.

According to the Bureau of Industry and Security [BUR 12], offset compensation can take different forms:

– Coproduction: transactions made under government-to-government agreements that allow the transfer of technology to a foreign company in order to build one piece or an entire defense article.

– Credit assistance: credit assistance comprises direct loans, brokered loans, loan guarantees, help in getting favorable payment conditions, credit extensions and lower interest rates. This type of offset excludes the use of credit that exceeds the demands of the contract and is allowed by the terms of the agreement, to be applied to future offsets compensation. The credit assistance is often classified as an indirect offset, but it can also be direct.

– Investment: an investment derived from either a direct or indirect offset agreement aiming to create or develop a company in the foreign country.

– Licensed production: production by external suppliers and domestic manufacturers of a component or item of a defense system under the mandate and based on technical information from the seller. This can take both a direct and an indirect form.

– Purchases: indirect transactions that involve the purchase of off-the-shelf items from the buyer (offset recipient).

– Subcontracts: production of a part or component of the purchased defense article by a foreign company. It does not require licenses or consent and is a direct agreement between the seller and the manufacturing company.

– Technology transfer: occuring as a result of a compensation agreement, and usually this takes the form of research and development, provision of technical assistance to the affiliate or joint venture of overseas investment, or other activities under the commercial agreement between the seller and the buyer.

– Training: covers training related to the production or maintenance of the exported defense product. The training, which can be a direct or indirect offset, may be required in various areas.

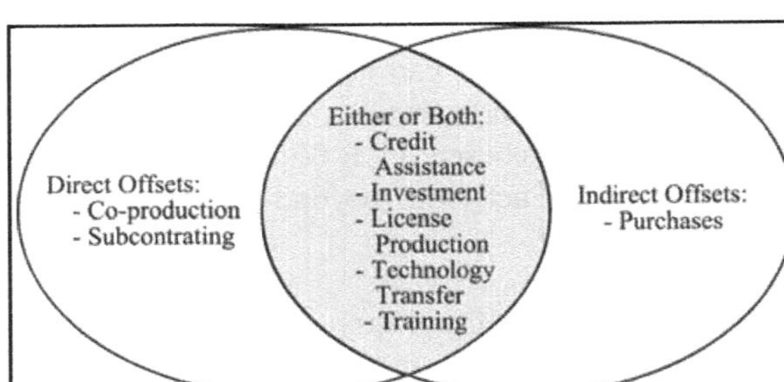

Figure 5.2. *Types of offsets [BUR 12, p. 27].*

We can infer that the various types of offset agreements can help in removing some of the barriers to innovation presented previously, and summarized in terms of cost factors, knowledge factors, market factors and institutional factors. Each type of offset can act against specific barriers:

– Credit assistance can fight against cost factors.

– Training, technology transfer, coproduction, subcontract and licensed production can act against knowledge factors.

– Purchases can help to solve problems regarding market factors.

– Investment can assist in overcoming institutional factors.

Nevertheless, we must remember that offsets can also be used in other ways. In other words, those relationships between barriers to innovation and corresponding forms of offsets do not exhaustively explore the potential of such transactions.

Similar definitions are used in Brazil [MIN 02]. The main difference could be found with respect to training. Training, as an offset benefit, is accepted in Brazilian context if it is related to the technology transfer and if it will represent a qualitative increase in the technological level of the country in the near future.

Nowadays, offset agreements in Brazil must include technology transfer because this is the way in which Brazil overcome the barrier that separates the developed countries from the non-developed countries in the shortest period.

5.4. How exactly is Brazil using offsets to overcome the major obstacles to innovation?

To better understand the current state of offset agreements in Brazil, it is important to briefly trace its history.

Brazil started using offsets before the Second World War. At that time, the Brazilian Government and Italy negotiated the purchase of Italian submarines in exchange of coffee and cotton [SCH 04]. By that time, the Brazilian Army (*Exército Brasileiro*) had already performed similar negotiations with the Germans by exchanging arms for cotton.

In the 1950s, the Brazilian Air Force (*Força Aérea Brasileira* – FAB) bought some Gloster Meteor airplanes from England and paid them the equivalent amount in cotton [ALV 08].

Until then, the Brazilian national industry has not benefitted much from offset agreements. The contracts were confined to the most commonly traded goods in an attempt to maintain a favorable trade balance.

In 1974, the extinct Ministry of the Air Force purchased some fighter airplanes, model F-5 Tiger, and had demanded offset compensation as part of the deal, namely the transfer of technology to be used in the advanced design of some airplane's pieces. This technology was transferred to EMBRAER.

In the 1980s, EMBRAER negotiated with an Italian company in the joint production of the AMX-fighter in Brazil, which represented a strong increase in its innovative power [TOL 93]. Strong evidence of the positive spillover is the EMBRAER's position in the Global Innovation 1000 rank [JAR 12]. EMBRAER is currently one of the most powerful small and medium airplane constructors in the world, and civilian customers are among its main clients. This is illustrative of how offsets (originally conceived for military purposes and necessities) can help enterprises to innovate and become leaders in their markets.

In addition, Brazil has successful offset cases in the aeronautical as well as in the naval industries. It is the only country in the Southern Hemisphere which is able to build and maintain conventional submarines, because of the latest agreements that oblige the sellers to construct submarines for the Brazilian Navy (*Marinha do Brasil*), in Brazil using Brazilian staff.

From then on, many other offset agreements have been signed. The latest agreements are also the most emblematic, due to the large sums involved in the transactions. One of the offsets is related to the acquisition of the Super Cougar helicopters, and the other offset concerns the Submarines Development Program (PROUB). They were both signed in December of 2008 and started in the beginning of 2009. Such contracts are equivalent to approximately EUR 6,130,000,000.00 in compensation, have a long-term time horizon and are being performed by French enterprises.

In the near future, the Brazilian Government plans to develop another successful offset. It involves the purchasing/manufacturing of the KC-390 project (a military cargo aircraft), and the participation of various countries, 90 countries, according to the Brazilian Air Force. All these projects are expected to improve training, increase the investment and facilitate access to state-of-art technologies that otherwise would not be available.

One of the most important issues for offset participants is where to focus. It is essential to keep in mind that the purchaser has to know his/her own necessities in order to prevent offset failure.

In Brazil, the most usual operations within offset agreements include training, technology transfer and investment. Offsets have been used to overcome the high costs of innovation in Brazil, the country's infrastructure deficiency and some significant technological barriers. Training and technology transfer are intended to prevent some lacunae in terms of knowledge and skills and to prompt the acquisition of core competences.

5.5. Some methodological considerations and empirical results

The purpose of this research is to answer the following question: can offset agreements contribute to the removal of technological barriers and be a trigger to innovation in the context of Brazil's Defense Industry? An electronic survey of Brazilian defense companies was conducted in November 2012 to provide some insight into the opportunities, challenges and constraints that those companies have been experiencing with offsets.

The research participants were asked about their perceptions concerning: (1) the major restrictions and barriers to innovation and (2) the perceived role of the offset agreements as a means to overcome such barriers and constraints, along with other innovation-related data. The reference period for the survey comprised the years 2009, 2010 and 2011, the same time frame used in other research that served as the basis for the present study, namely the "Oslo Manual" [ORG 05] and the PINTEC – 2012 [IBG 12]. The latest is a survey about innovation developed by a Brazilian Governmental agency, the Brazilian Institute of Geographic and Statistics (IBGE).

5.5.1. *Measuring instrument*

The original online questionnaire is divided into nine parts: the company's characterization (part 1), the respondent's characterization (part 2), product innovation (part 3), process innovation (part 4), innovation activities (part 5), cooperation (part 6), government support (part 7), barriers to innovation (part 8) and offset agreements (part 9). The survey follows the Oslo Manual guidelines [ORG 05] regarding parts 3, 4, 5 and 8 and the PINTEC guidelines [IBG 12] regarding parts 6 and 7 in order to provide results capable of future comparison with other research. It is important to state that PINTEC also follows the Oslo Manual guidelines. Parts 1, 2 and 9 are newly constructed in order to meet the research objectives.

A pilot study was conducted with Brazilian specialists on innovation. A pretest was made with 18 defense companies. Some contributions that originated from the pretest and the pilot study were incorporated in the survey.

The present chapter is based on parts 1, 2, 3, 4, 8 and 9 of the original questionnaire. Parts 5 (innovation activities), 6 (cooperation) and 7 (government support) were not used in the present chapter. Part 1 was constituted by multiple-choice questions related to the companies' characterization, namely their position in Brazil, how many years they have existed, how many employees they employ, their main activity and their target markets. Part 2 was also constituted by multiple-choice questions intended to characterize the respondent, such as their age, number of years in the company, their position in the company and educational level. Part 3 was composed of multiple-choice questions to discover if the company innovated with regard to products and the quantity of implemented innovations. It used check box questions to discover the innovation level and also to know how, where and by whom the innovation was made. Part 4 had the same design as Part 3, but it was directed toward innovations on process. Part 8 was composed of a grid question. Respondents had to answer using a three-point Likert-type scale (low, medium and high), to analyze the intensity of barriers. Part 9 used multiple-choice questions to gain a deeper understanding of the reasons to participate in offsets. It also aimed to gather information on how extensive their participation was and their intention to participate in future offsets. It also included a grid question with the same items used in Part 8, but intended to evaluate the companies' belief about the ability of the offset to remove barriers to innovation.

5.5.2. *A brief description of the companies surveyed and the respondents*

The research comprises member companies of the Brazilian Defense and Security Industries Association (*Associação Brasileira das Indústrias de Materiais de Defesa e Segurança* – ABIMDE). When the online survey was conducted, the association was composed of 172 civilian companies, but only 147 were considered eligible to participate in the survey because 25 did not meet the research's requirements. The survey uses the company as a unit of analysis, and the respondent was the main member responsible for the innovation area. A total of 54 companies have completed the online questionnaire, representing a response rate of 37%. Nine questionnaires that were returned were not used in the statistical analysis because the companies did not develop innovative projects or were not formally constituted between 2009 and 2011.

Although most companies were located in São Paulo (69%), companies from Rio de Janeiro (13%), Minas Gerais (7%), Paraná (5%), Ceará (2%),

Rio Grande do Sul (2%) and Santa Catarina (2%) have also returned the questionnaire.

A total of 69% of these companies had been established more than 10 years ago, 22% between 3 and 10 years and 9% quite recently, since they have existed for less than 3 years.

Twenty-five companies (56%) employ more than 49 employees, five companies (11%) had between 50 and 99 employees, 10 companies (22%) had between 100 and 499 employees and five other companies (11%) exceed 500 employees.

Twenty-five companies (56%) were industrial, 19 companies (42%) provide trading services and one company (2%) was responsible for trading products.

Forty-one companies (91%) were owned exclusively by national shareholders. The target market of 64% of companies is the domestic market, 29% operate within the domestic and international markets and only 7% concentrate exclusively on the international market.

A total of 51% of companies provide goods and services to the public and private sectors, 27% only to the public sector and 22% exclusively to the private sector.

It is worth mentioning that most member companies of ABIMDE provide goods and services to military and civilian institutions, as shown in Figure 5.3.

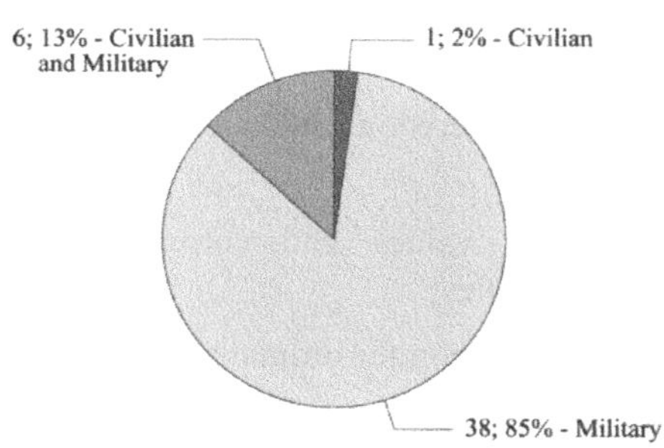

Figure 5.3. *Recipient of goods and services provided*

This finding illustrates the existence of dual-technology application, that is the use of products and services provided by defense companies in both civilian and military spheres. This means that technology obtained through a purchase agreement of military equipment can benefit civilian industry.

A total of 67% of the respondents were more than 41 years old, 40% had worked for more than 10 years in the company, 87% (Figure 5.4) were directors or managers and 58% had obtained some academic degree after completing an undergraduate course (specialization, master's or/and doctoral degree).

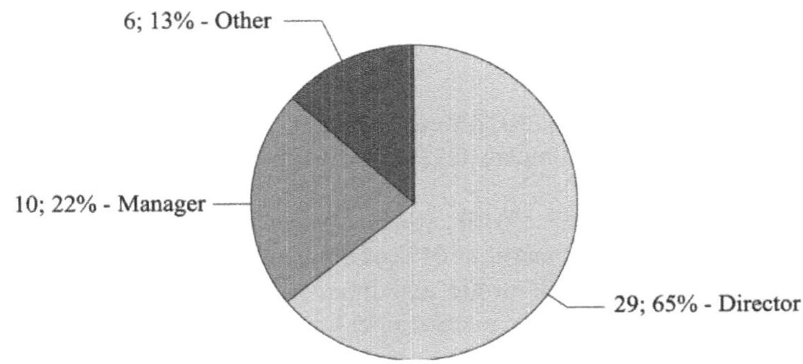

Figure 5.4. Respondents by position in the company

5.5.3. *Some empirical evidences on the perceived role of offsets*

The data collected through the online survey were coded and subjected to statistical analysis, using the Statistical Package for the Social Sciences (SPSS) version 20.0.

Some descriptive and inferential statistics were calculated as follows [FIE 09]:

– frequencies (absolute values and corresponding percentages) for categorical or nominal variables;

– mean and standard deviation for ordinal and interval variables;

– non-parametric Chi-square (χ^2) test, Kendall Tau (τ) test or Mann–Whitney (U) test were calculated to assess the relationships between variables. As suggested by the literature [FIE 09], the Chi-square test was used to analyze the association between two categorical variables, and Kendall Tau (τ) was used to evaluate the relationship between two ordinal variables. Mann–Whitney (U) test was used to measure the differences between variables, when the dependent variable is ordinal or interval, but not normally distributed, when there is, or there is not, homogeneity of variance, and the independent variable is categorical. The 0.05 level of significance was used when interpreting the results.

5.5.4. *Main research questions and results*

To know if the offset agreements were fulfilling their obligations to increase innovation and combat the barriers to innovation, four main questions, that will be answered in the subsequent sections, were posed.

Before proceeding with the main questions and empirical evidence, it is necessary to provide some general information concerning the surveyed companies' participation in offsets. According to the data gathered, five out of the 45 companies have participated in offsets before 2009 or after 2011; there has been a slight increase, from five to seven, between 2009 and 2011. Five companies out of those seven have received compensation as a result of the offset contract, and reported at least one innovation during that period as a consequence of an offset agreement.

5.5.4.1. *Have companies that participated in offset agreements become more innovative?*

It is worth noting that only product innovations or process innovations are associated with technological innovation [AFU 98, CAR 11]. Bearing this in mind, 87% of the companies reported having introduced at least one product innovation, 51% of the companies reported having introduced at least one process innovation and only 13% (six companies) did not innovate in the period. These findings draw attention to the fact that all companies who innovated it process have also introduced innovative products, probably because the ability to introduce new products in the market should be supported by the production processes [DAM 01].

To test for differences in the amount of new (or significantly improved) products and processes, introduced in the market between 2009 and 2011, among companies that participated in offset agreements and those that did not, we have performed the Mann–Whitney (U) test. The number of products and innovative processes developed were used as dependent variables, and participation in compensation arrangements was considered as the independent variable.

Table 5.2 presents the descriptive statistics and the Mann–Whitney test for the number of new or significantly improved products (goods and/or services) introduced as a result of the companies' participation in compensation arrangements in two periods: (1) between 2009 and 2011 and (2) the period before 2009 or after 2011.

Participation in offset agreements	*N*	Average innovative products (SD)	Z	Significance
Between 2009 and 2011				
Yes	6	3.67(1.86)	–0.04	0.97
No	33	3.73(1.94)		
Before 2009 or after 2011				
Yes	5	3.00(1.73)	–0.89	0.37
No	34	3.82(1.93)		

Table 5.2. *Number of new or significantly improved products (goods and/or services) introduced due to participation in offset agreements in two different time periods*

There were no differences in the number of innovative products developed by the companies that participated in compensation arrangements and those that did not participate in any of the reference periods ($p > 0.05$).

Table 5.3 presents the descriptive statistics indicators and results of Mann–Whitney test for the number of new or significantly improved processes introduced between 2009 and 2011, following the companies' participation in offset agreements in two time periods: (1) between 2009 and 2011 and (2) the period before 2009 or after 2011.

We have found statistically significant differences in terms of the average number of innovative processes among the companies that participated in offset agreements between 2009 and 2011 and those that did not (see Table 5.3). When it comes to innovative processes, the results show that companies that participated in offset agreements between 2009 and 2011 present an average of 5.25 innovative processes while those that did not participate present, on average, 2.95 cases in the same period.

Participation in offset agreements	*N*	Average innovative processes (SD)	Z	Significance
Between 2009 and 2011				
Yes	4	5.25(1.50)	–0.02	0.04
No	19	2.95(1.99)		
Before 2009 or after 2011				
Yes	2	6.00(0.00)	–1.70	0.09
No	21	3.10(1.99)		

Table 5.3. *Number of processes new or significantly improved introduced due to participation in offset agreements in two different time periods*

We have not found statistically significant differences between the two groups of companies in the other periods – before 2009 or after 2011 (Table 5.3).

5.5.4.2. *What were the main barriers to innovation faced by companies between the years 2009 and 2011?*

This question was necessary in order to answer the subsequent question. To know if the offset agreements are capable of combatting the barriers to innovation, first it is required to know what barriers companies faced.

To find an answer to this question, it is necessary to know that 25 companies that responded to the survey (55%) reported some sort of barrier to innovation.

Table 5.4 presents the barriers to innovation that were classified by companies as having high or medium relevance.

The most relevant barriers to innovation identified by the surveyed companies were the same found by IBGE [IBG 12] in the studies conducted from 2000 to 2008 (PINTEC).

5.5.4.3. *Do the companies believe that offset agreements can help in overcoming barriers to innovation?*

One fundamental fact is that 36 companies out of the 45, representing 80% of the respondents, reported that offset agreements can help to fight barriers to innovation. These respondents believe that offset can combat different barriers to innovation with high or medium efficacy as shown in Table 5.5.

Barrier to innovation	Companies (% of the total)
High costs to innovate	100
Lack of funding sources	87
Excessive economic risks	83
Lack of qualified personnel	73
Scarcity of opportunities for cooperation	65
Organizational rigidity	67
Weak consumer response	61
Lack of information on markets	57
Shortage of adequate technical services	60
Difficulty in adequacy to norms and standards	50
Lack of information on technology	38
Innovation's centralization activities in other company's group	14

Table 5.4. *Percentage of companies that classified the presented barriers to innovation as having high or medium relevance*

Barrier to innovation	Companies (% of the total)
High costs to innovate	84
Excessive economic risks	62
Lack of funding sources	59
Lack of qualified personnel	56

Table 5.5. *Percentage of companies that believe offset agreements can help in overcoming barriers to innovation with high or medium efficacy*

When comparing and contrasting these results with those previously presented in Table 5.4, we can infer that the companies' belief in offset efficacy is connected with the barriers most frequently encountered (Table 5.5).

To evaluate the relationship between the relevance attributed to the encountered barriers to innovation and the belief in the offset agreements to overcome them, the correlation of Kendall Tau test was computed. Table 5.6 presents the descriptive statistical indicators, as well as the test results of the Kendall Tau test.

Relevance of barriers to innovation	Offset agreements ability to fight the barriers to innovation			τ	Significance
	Low	Medium	High		
	n(%)	n(%)	n(%)		
Excessive economic risks					
Low	3(75%)	1(25%)	0	0.59	0.004
Medium	0	5(62.5%)	3(37.5%)		
High	0	3(33.3%)	6(66.7%)		
High costs to innovation					
Low	0	0	0	0.22	0.32
Medium	0	3(42.9%)	4(57.1%)		
High	0	3(21.4%)	11(78.6%)		
Lack of funding sources					
Low	3(100%)	0	0	0.60	0.005
Medium	1(20%)	3(60%)	1(20%)		
High	1(9.1%)	3(27.3%)	7(63.6%)		
Organizational rigidity					
Low	5(100%)	0	0	0.68	0.005
Medium	1(25%)	2(50%)	1(25%)		
High	1(16.7%)	1(16.7%)	4(66.7%)		
Lack of qualified personnel					
Low	5(83.3%)	0	1(16.7%)	0.55	0.01
Medium	2(20%)	6(60%)	2(20%)		
High	0	1(33.3%)	2(66.7%)		
Lack of information on technology					
Low	4(40%)	6(60%)	0	0.68	0.002
Medium	0	4(66.7%)	2(33.3%)		
High	0	0	2(100%)		
Lack of information on market					
Low	7(77.8%)	1(11.1%)	1(11.1%)	0.61	0.003
Medium	3(42.9%)	3(42.9%)	1(14.3%)		
High	0	0	4(100%)		
Scarcity of opportunities for cooperation					
Low	0	4(66.7%)	2(33.3%)	0.27	0.22
Medium	1(16.7%)	2(33.3%)	3(50%)		
High	0	2(28.6%)	5(71.4%)		
Difficulty in adequacy to norms and standards					
Low	5(55.6%)	3(33.3%)	1(11.1%)	0.39	0.08

Table 5.6. *Relationship between barriers to innovation and belief in offset agreement's efficacy in overcoming them*

Medium	0	1(50%)	1(50%)		
High	2(28.6%)	1(14.3%)	4(57.1%)		
Weak consumer response					
Low	7(87.5%)	1(12.5%)	0		
Medium	0	1(25%)	3(75%)	0.51	0.02
High	2(33.3%)	1(16.7%)	3(50%)		
Shortage of adequate technical services					
Low	4(66.7%)	2(33.3%)	0		
Medium	3(60%)	1(20%)	1(20%)	0.60	0.01
High	0	1(20%)	4(80%)		
Innovation's centralization activities in another company's group					
Low	6(66.7%)	2(22.2%)	1(11.1%)		
Medium	0	0	0	0.63	0.04
High	0	0	2(100%)		

Table 5.6. *(Continued) Relationship between barriers to innovation and belief in offset agreement's efficacy in overcoming them*

Statistically significant correlations were found between the offset efficacy and the following barriers to innovation: excessive economic risks, lack of funding sources, organizational rigidity, lack of qualified personnel, lack of information on technology, lack of information on market, weak consumer response, shortage of adequate technical services and innovation's centralization activities in other company's group. The correlation coefficients indicate positive and moderate associations, ranging from 0.51 ("weak consumer response") to 0.68 ("organizational rigidity" and "lack of information on technology"). The positive correlation coefficients mean that the offset agreements are perceived as performing an effective role in overcoming those barriers to innovation.

Empirical evidence that gives support to this argument is that, when asked about their interest in participating in an offset agreement in the future, 78% (35 out of the 45) of the companies responded positively.

5.5.4.4. *Can offsets help in overcoming barriers to innovation?*

To test the possible existence of a relationship between participation in offset agreements and the intensity of innovation, that is the extent to which they hamper barriers to innovation, Chi-square test and Fisher's exact test were performed. The relevance assigned to each barrier to innovation was taken as the dependent variable and participation in compensation arrangements as the independent variable.

Table 5.7 presents the results of the descriptive statistics and the Fischer's exact test to explore the relevance assigned to each barrier to innovation faced by companies on their participation in offset agreements.

Relevance of barriers to innovation	Participation in offset agreements		Fischer's exact test	Significance
	Yes	No		
	n(%)	*n*(%)		
Excessive economic risks				
Low	1(33.3%)	3(14.3%)	3.766	0.14
Medium	2(66.7%)	6(28.6%)		
High	0	12(57.1%)		
High costs to innovate				
Low	0	0		
Medium	1(33.3%)	6(28.6%)	0.028	1.00
High	2(66.7%)	15(71.4%)		
Lack of funding sources				
Low	0	3(15%)		
Medium	3(100%)	4(20%)	6.051	0.03
High	0	13(65%)		
Organizational rigidity				
Low	2(66.7%)	5(27.8%)		
Medium	1(33.3%)	7(38.9%)	1.890	0.60
High	0	6(33.3%)		
Lack of qualified staff				
Low	2(66.7%)	4(21%)	2.324	0.24
Medium	1(33.3%)	11(58%)		
High	0	4(21%)		
Lack of information on technology				
Low	3(100%)	10(55.6%)		
Medium	0	6(33.3%)	1.593	0.65
High	0	2(11.1%)		
Lack of information on market				
Low	2(66.7%)	8(40%)	0.878	1.00
Medium	1(33.3%)	8(40%)		
High	0	4(20%)		

Table 5.7. *Relevance of encountered barriers to innovation and companies' participation in offset agreements*

Scarcity of opportunities for cooperation				
Low	1(33.3%)	7(35%)	0.457	1.00
Medium	1(33.3%)	6(30%)		
High	1(33.3%)	7(35%)		
Difficulty in adequacy norms and standards				
Low	1(33.3%)	10(52.6%)	1.182	0.75
Medium	1(33.3%)	3(15.8%)		
High	1(33.3%)	6(31.6%)		
Weak consumer response				
Low	2(66.7%)	7(35%)	1.624	0.75
Medium	0	7(35%)		
High	1(33.3%)	6(30%)		
Shortage of adequate technical services				
Low	2(100%)	6(33.3%)	35%	0.31
Medium	0	7(38.9%)		
High	0	5(27.8%)		
Innovation's centralization activities in other company's group				
Low	1(100%)	11(84.6%)	0.167	1.00
Medium	0	0		
High	0	2(15.4%)		

Table 5.7. *(Continued) Relevance of encountered barriers to innovation and companies' participation in offset agreements*

There were no statistically significant results for the intensity of barriers to innovation and participation in offset agreements, except for "lack of funding sources".

5.6. Conclusion

Innovation is widely recognized as a trigger for growth at organizational, national and international levels, especially in the present context of globalization and unprecedented level of competitiveness. A large and growing body of literature has investigated the innovation process, the most effective policies and tools to achieve innovation, and the major impediments toward innovation. In this chapter, we looked specifically at offsets as a vehicle for removing some barriers to innovation.

Over the past decades, offsets have attracted a great deal of attention from governments and companies all over the world, particularly in the developing economies. Many studies pinpoint their beneficial effects, such as the strengthening of power to develop national skills and know-how, and their potential in stimulating domestic industries and promoting economic development.

Brazil has been using offsets to overcome technological barriers to innovation since the 1970s. As a result, the country has witnessed the emergence and development of civilian industries that have absorbed and capitalized on investments and technology transfers. The benefits of offsets in the innovation processes of Brazilian enterprises can be supported by several stories. The aircraft manufacturer EMBRAER is one of the best illustrations of the benefits provided by offsetting. AEL Sistemas and the Brazilian Navy Yard (*Arsenal de Marinha do Rio de Janeiro*) are also other good examples of what can be achieved through such contracts or agreements.

This study has used an online questionnaire to survey the civilian companies, members of the Brazilian Defense Industry Association, about various issues on innovation, with a special emphasis on offsets. According to the survey, the offset agreements were significantly associated with the existence of process innovation in the surveyed companies in the period 2009–2011. No relationship was found between the participation of enterprises in offset agreements and the occurrence of product innovations. General findings derived from questions posed to respondents suggest that the companies that participated in offset agreements are not more innovative than those that did not participate in offsetting.

However, it is important to note that all companies who reported innovation in processes also reported product innovation. Five companies that received some kind of compensation during the period 2009–2011 reported having implemented at least one innovation as a result of a compensation agreement.

The main barriers to innovation experienced by companies in the survey are the same that had already been identified by Brazilian companies in the period 2000–2008. This consistent finding should call for government intervention because three of the barriers cited are economic (high costs of innovation, excessive economic risk and lack of funding sources), and only one barrier is structural and inherent to each company (lack of qualified personnel). The latest – "lack of qualified staff" – reflects a wider problem of educational policy that also calls for government intervention. Being aware of this problem, Brazilian politicians have launched the program "*Ciência sem Fronteiras*" in 2011, aimed at increasing the level of education, knowledge and skills of the population.

Companies that have experienced barriers to innovation perceive the offset agreements as an effective tool in overcoming some major impediments to innovation. This is true in all aspects identified and with the same intensity in which barriers occur. This is probably the reason why they show interest in participating in future offset agreements. A total of 87% of companies consider participating in an offset agreement as an investment, not a cost. This understanding of offsets is of extreme importance because it highlights the positive nature of such contracts instead of the negativity that permeates some discourses.

Empirical evidence that is worth mentioning is that, among the companies participating in this survey, the only barrier to innovation that apparently could be transposed through compensation arrangements was the "lack of funding sources".

As compensation arrangements involve long-term contracts and the benefits of recent agreements (signed in December 2008 and started in 2009) will only occur in the near future, it is expected that future surveys will find a greater significance among the participation in offset arrangements and the intensity of barriers to innovation experimented with by the participant companies.

5.7. Bibliography

[AFU 98] AFUAH A., Innovation Management, University Press, Oxford, 1998.

[ALV 08] ALVES F.A.S., Acordos de Compensação – Offset – Firmados pelas Forças Singulares: Melhorias a realizar, Escola de Comando e Estado Maior do Exército, Rio de Janeiro, 2008.

[AMA 90] AMARA R., "New directions for innovation", *Journal Futures*, vol. 22, no. 2, pp. 142–152, 1990.

[BRA 04] BRAUER J., DUNNE J.P. (eds), *Arms Trade and Economic Development*, Routledge, London, 2004.

[BRA 11] BRASIL, Lei no. 12.545/2011, Imprensa Nacional, Brasília, 2011.

[BRA 12] BRASIL, Lei no. 12.598/2012, Imprensa nacional, Brasília, 2012.

[BRE 98] BRENNAN M., "Government pro-active countertrade: a decade of deals", Working paper no. 171, Murdoch University, Perth, 1998.

[BUR 12] BUREAU OF INDUSTRY AND SECURITY, *Offset in Defense Trade – Sixteenth Study*, Bureau of Industry and Security, Washington, DC, 2012.

[CAR 11] CARVALHO J.E., LOPES J.A.A., REIMÃO C.M., *Inovação, Decisão e Ética*, Sílabo, Lisbon, 2011.

[DAM 01] DAMANDOPOUR F., GOPALAKRISHINAN S., "The dynamics of the adoption of product and process innovations in organizations", *Journal of Management Studies*, vol. 38, no. 1, pp. 45–65, 2001.

[DUT 12] DUTTA S., *The Global Innovation Index 2012*, Stronger Innovation Linkages for Global Growth, INSEAD, Fontainebleau, 2012.

[EUR 09] EUROPEAN UNION COUNCIL, Council Regulation (EC) No. 428/2009, Brussels, 2009.

[FIE 09] FIELD A., *Discovering Statistics Using SPSS*, Sage Publications, London, 2009.

[FRE 00] FREIRE A., *Inovação*, Verbo, Lisbon, 2000.

[GUP 08] GUPTA P., *Inovação Empresarial no Século XXI.*, Vida Económica, Porto, 2008.

[HOW 10] HOWSE R., "Beyond the countertrade taboo: why the WTO should take another look at barter and countertrade", *University of Toronto Law Journal*, vol. 60, no. 2, pp. 289–314, 2010.

[IBG 00] IBGE, *Pesquisa de Inovação Tecnológica*, IBGE, Rio de Janeiro, 2000.

[IBG 03] IBGE, *Pesquisa de Inovação Tecnológica*, IBGE, Rio de Janeiro, 2003.

[IBG 05] IBGE, *Pesquisa de Inovação Tecnológica*, IBGE, Rio de Janeiro, 2005.

[IBG 08] IBGE, *Pesquisa de Inovação Tecnológica*, IBGE, Rio de Janeiro, 2008.

[IBG 12] IBGE, *Pesquisa de Inovação Tecnológica*, IBGE, Rio de Janeiro, 2012.

[IBM 06] IBM, *Expanding the Innovation Horizon: The Global CEO Study*, IBM Corporation, Somers, NY, 2006.

[IVO 04] IVO R.C., A Prática do Offset como instrumento dinamizador do desenvolvimento industrial e tecnológico, Universidade de Brasília, Centro de Desenvolvimento Sustentável, Brasília, Brazil, 2004.

[JAR 12] JARUZELSKI B., LOEHR J., HOLMAN R., *The Global Innovation 1000: Making Ideas Work*, Booz & Company, New York, 2012.

[JOB 11] JOBS S., "O legado depois de Jobs", by SALVADO L., *Executive Digest*, vol. 67, no. 2, 2011.

[JON 02] JONES W.E., 2002. "The value of military industrial offsets", *Defense Institute of Security Assistance Management, Journal of International Security Assistance Management*, vol. 24, no. 2, pp. 108–118, 2001–2002.

[LÖÖ 05] LÖÖF H., HESMATI A., The impact of public funding on private R&D investment: new evidence from a firm level innovation study, Working paper 06, The Royal Institute of Technology, Centre of excellence for studies in Science and Innovation, Stockholm, 2005.

[MAD 09] MADRID-GUIJARRO A., GARCIA D., VAN AUKEN H., "Barriers to innovation among Spanish manufacturing SMEs", *Journal of Small Business Management*, vol. 47, no. 4, pp. 465–488, 2009.

[MIN 02] MINISTÉRIO DA DEFESA, Política e as Diretrizes de Compensação Comercial, Industrial e Tecnológica do Ministério da Defesa, Anexo II, Ministério da Defesa, Brasília, Brazil, 2002.

[MIN 11] MINISTRY OF DEFENSE, Defense Procurement Procedure, Government of India, 2011.

[MIN 12] MINISTÉRIO DA CIÊNCIA TECNOLOGIA E INOVAÇÃO, Distribuição percentual dos dispêndios nacionais em pesquisa e desenvolvimento (P&D), segundo setor de financiamento, países selecionados, 2000–2010, available at http://www.mcti.gov.br/index.php/content/view/336709/Distribuicao_percentual_dos_dispendios_nacionais_em_pesquisa_e_desenvolvimento_P_D_segundo_setor_de_financiamento_paises_selecionados.html, 2012.

[MOD 04] MODESTI A., AZEVEDO A.E.M., "Offset: teoria e prática", *Panorama da Prática do Offset no Brasil – uma visão da negociação internacional de Acordos de Compensação Comercial, Industrial e Tecnológica*, Suspensa, Zuhair Warwar, Brasília, Brazil, 2004.

[NAY 91] NAYAK P.R., *Managing Rapid Technological Development*, Arthur D. Litlle, Boston, MA, 1991.

[NEU 85] NEUMAN S.G., *Coproduction, Barter and Countertrade: Offsets in the International Arms Market*, ORBIS, pp. 167–181, 1985.

[ORG 63] ORGANIZATION FOR ECONOMIC CO-OPERATION AND DEVELOPMENT (OECD), *Frascati Manual,* 1st ed., OECD, Paris, 1963.

[ORG 05] ORGANIZATION FOR ECONOMIC CO-OPERATION AND DEVELOPMENT (OECD), *Oslo Manual*, 3rd ed., OECD, Paris, 2005.

[ORG 12a] ORGANIZATION FOR ECONOMIC CO-OPERATION AND DEVELOPMENT (OECD), *Education at a Glance: OECD Indicators 2012*, OECD, Paris, 2012.

[ORG 12b] ORGANIZATION FOR ECONOMIC CO-OPERATION AND DEVELOPMENT (OECD), *OECD Science, Technology and Industry Outlook 2012*, OECD, Paris, 2012.

[OWE 11] OWENS D.A., *Creative People Must Be Stopped: 6 Ways We Kill Innovation (Without Even Trying)*, Jossey Bass, San Francisco, CA, 2011.

[PAR 06] PARGAC P., *What Issues Does the Czech Republic Face Concerning Offsets in Context of Military Purchase*, Naval Postgraduated School, Monterey, CA, 2006.

[SAR 10] SARKAR S., *Empreendedorismo e Inovação,* 2nd ed., Escolar, Lisbon, 2010.

[SCH 00] SCHACHT W.H., Industrial competitiveness and technological advancement: debate over government policy, CRS Report RL33528, Congress Research Service, Washington, DC, 2000.

[SCH 04] SCHNAIDERMAN B., *Guerra em Surdina*, Cosac Naify, São Paulo, 2004.

[SUM 10] SUMAN M., Defence offsets: proving detrimental to the services", available at www.indiandefensereview.com/news/defense-offsets-proving-detrimental-to-the-services/

[TOL 93] TOLENTINO P.E.E., *Technological Innovation and Third World Multinational*, Routledge, London, 1993.

[UND 11] UNDR, Human Development Report 2011, Palgrave, New York, 2011.

[UNE 11] UNESCO, Education for All Global Monitoring Report, UNESCO, Paris, 2011.

[VER 09] VERMA S., "Offset contracts under defense procurement regulations in India: evolution, challenges and prospects", *Journal of Contract Management*, vol. 7, pp. 17–32, 2009.

Chapter 6

Knowledge Management Fostering Innovation: Balancing Practices and Enabling Contexts

In this chapter, knowledge management (KM) is seen as an important driver for enhancing innovation performance levels. It is argued that a better understanding of knowledge processes antecedents and consequences is crucial for the improvement of management decisions concerning fostering innovation throughout KM initiatives. The chapter presents a theoretical framework to support the implementation of knowledge processes oriented towards innovation, highlighting the key elements that should be included in it.

6.1. Introduction

In recent years, there has been a widespread debate about the relevance of intangible assets in general and knowledge in particular. Indeed, there is now a broad consensus around knowledge being positioned as central to a sustainable competitive advantage.

Chapter written by Maria João SANTOS and Raky WANE.

In the mid-1990s, interest in the knowledge management (KM) theme mushroomed among academics, politicians, consultants and managers. Studies carried out by consultants such as KPMG and McKinsey suggested that various organizations were setting about implementing structured KM programs. There is no simple and consensual definition of KM capable of being applied to all organizations. Davenport and Prusak [DAV 98] consider KM consists of a set of processes and means for creating, utilizing and disseminating knowledge throughout the organization. In turn, Allee [ALL 03] affirms that KM practice should prioritize and support processes designed to create, maintain, share and renew organizational knowledge so as to boost the potential for economic growth whether through creating value or improving performance levels. More recently, Hislop [HIS 09] deems KM to be a broad-reaching term incorporating any deliberate effort seeking to manage the knowledge of employees in a particular organization. This may be achieved through a broad variety of methodologies, including those based on the use of purpose-designed technologies and those seeking to bring about the more efficient management of social processes, organizational structures, cultures and personalized management practices.

These different positions demonstrate that there are multiple and diverse, when not contradictory, interpretations as to just how KM should be put into effect (e.g. [GAO 08]). Indeed, the focus on KM has progressively shifted moving on from concentrating on the "hard" factors, based on technology and the reutilization of knowledge, to take on a "soft" perspective based on the prevailing interactions and social contexts (e.g. [MCE 00, CHO 03, GAO 08, HIS 09]). Narrower approaches to KM typically fail to appreciate the additional factors that impact knowledge processes and activities. McElroy [MCE 00] defends how the soft perspective leads to better organizational results as, beyond distributing and utilizing the existing knowledge as efficiently as under the hard approach, such methods also foster organizational learning and innovation through constant stimuli to discover and create new knowledge. Similarly, other authors (e.g. [NON 94, MAS 02]) posited that balancing and reconciling the hard and soft perspectives leads to the most innovative performances.

Although KM has become a central topic, in some cases the best efforts of organizations to manage knowledge have not achieved their objectives (e.g. [DE 00, LEE 03]). The question is no longer whether to manage organizational knowledge, but how to manage it to reach higher performances. In this chapter, we analyze the contribution made by KM practices to the innovation capacities, particularly in relation to product innovation. "A product innovation is the introduction of a good or service that is new or significantly improved with respect to its characteristics or intended uses. This includes significant improvements in technical specifications, components and materials, incorporated software, user friendliness or other functional characteristics" [ORG 05, p. 48].

The relationship between KM and innovation is conceptually supported in the literature from a range of research fields. The main theoretical contributions came from theories of strategic management (for instance resource-based view (RBV)), KM, intellectual capital (IC) and social capital. On the whole, fundamental conceptualizations and approaches reflect the assumption that knowledge assets are crucial to improve organizational performance, resulting in sustainable competitive advantages. Empirical evidence indicated that the implementation of KM practices, backed by a favorable organizational context, improves overall organization performance through higher profits, greater competitiveness and higher innovation performances (e.g. [FOR 02, CHO 03, ZAC 09]).

Different analyses have been enforced to address the interrelationship between KM and innovation. The empirical studies carried out may be grouped into two main categories: (1) the relationship between KM practices and performance levels and (2) the relationship between KM enablers, KM practices and performance levels. At this stage, we should comment on certain aspects of the various studies. First, although the relationship between enabling factors, practices and performance has been subject to analysis, very few studies have been based on models integrating these variables. Among those who opted for integrated models, the conceptual integration was carried out based upon the respective different approaches: systemic models [LEE 03], strategic – the "7-S" of McKinsey – [FOR 02] and process re-engineering [MAS 02]. Second, not all phases have been subject to equal levels of analysis and resulting in the actual effects of interactions between KM practices remaining unclear. Birkinshaw and Sheehan [BIR 02] state that no company can truly be active throughout all phases of the knowledge lifecycle. Nevertheless, analysis of the literature does encounter references pointing to how integrated cycle implementation does bring benefits to the organization (e.g. [CEN 04a]). The effects of interaction between creating and sharing knowledge based upon external alliances have also already been empirically verified [JIA 08] even though such is not the case with the remaining KM practices (identification and storage). Third, the relationship between the creation of knowledge and performance has turned up some odd results. For instance, in the model proposed by Choi and Lee [CHO 03], the creation of knowledge positively contributed toward intermediate outputs (creativity) that in turn contributed toward organizational performance (that also included innovation capacities). This and other similar empirical evidence suggests that, despite having been one of the most highly researched facets (e.g. [MAS 02, LEE 03, CAL 04, JIA 08]), analysis of the direct effects of knowledge creation on innovation performance levels may require greater depth and justification. Finally, measuring performance proves to be a fairly complex exercise that does not attract much consensus. Product innovations involve significant changes in the capabilities of goods or services. Both entirely new goods and services and significant improvements to existing products are included [ORG 05]. The empirical studies analyzed include both qualitative and quantitative metrics to measure innovation performance.

Despite the scientific evidence, there still remains a great deal of dissatisfaction as regard the actual leveraging of innovation. In its annual questionnaire, Boston Consulting Group (BCG) finds only 55% of executives express satisfaction as to the financial returns on investments in innovation [BOS 10], which would certainly seem to suggest that there is still much to understand about the potential means of leveraging company innovation capacities. A better understanding of how KM initiatives can be turned into value creation with positive impacts on innovation performance is fundamental to support management decisions and redirect efforts and investments toward those KM practices able to guarantee better results. How then can KM be managed to enhance innovative performances? The remainder of this chapter presents an integrated approach to answer this question, highlighting the relationships between KM enablers, KM practices and innovative performances.

6.2. Knowledge management and innovation

Based on the RBV and knowledge-based view (KBV), which is a recent extension of the first, some authors argued that valuable and unique knowledge, among other capabilities, leads to higher organizational and innovative performances. Consistent with this, authors such as Nonaka *et al.* [NON 00], Pralahad and Venkatram [PRA 03], and Bogner and Bansal [BOG 07] deem the learning processes underlying creating knowledge to be the very foundation stone of innovation, which, in turn, reflects in higher levels of organizational performance.

To leverage value from knowledge creation and using organizations the focus should be on implementing an effective KM that requires management of knowledge processes and implementation of specific practices designed to broaden organizational knowledge. According to Darroch [DAR 03, DAR 05], KM comprises three components – acquisition, dissemination and responsiveness (or use) – all of which positively affect innovation activity. In order for innovation to occur, organizations first need to have knowledge about the internal and external forces that affect the firm. Second, knowledge must flow freely around the firm so as to ensure that people within levels and departments of the organization are exposed to new knowledge that interacts with the knowledge already held. Finally, knowledge gained from multiple sources should apply, which means that the organization should respond to the various types of knowledge it has access to. The author points out that the highest KM impacts were identified for incremental innovation, that is improving or revising existing products or services.

Several other pieces of empirical evidence support that KM has a positive impact on innovative performances. In general, study results indicated that knowledge creation (e.g. [LEE 03, CAL 04, JIA 08]), dissemination and usage (e.g. [MCA 00, BIR 02]) are sources of innovation and positively impact company innovation performance levels, in terms of new products (NP), productivity, quality and time-to-market (e.g. [MAS 02]). Furthermore, contrasting theoretical contributions with empirical results, we find broad consensus on three core factors for the effective management of knowledge: enablers relating to a knowledge friendly organizational environment, specific KM management/production processes and organizational results.

The identification of these key components is neither new nor distinctive. However, there is a lack of study on the integrated interrelationship between these three factors (some exceptions are [FOR 02, MAS 02] and [LEE 03]). Additionally, even though many authors posited this correlation between KM and innovation, a large majority of them focused their analysis on practices design to create and share knowledge, excluding the other KM processes. For this reason, there is a clear need to reach out to the other phases inherent to the KM cycle and integrate the findings into the overall variables making up KM.

If the successful implementation of KM to foster innovation depends on its antecedents (organizational context) and consequents (innovative performance), then we must look at these components from an integrated perspective. Following this argument, leveraging KM practices to achieve innovation goals requires a deep understanding of the knowledge processes, what critical issues influence such implementation and how all of these factors relate to each other. As a result, we posit that systems thinking provides a foundation that can facilitate such an integrative understanding and can enhance an effective KM practice. Systems thinking theory is a conceptual framework for problem solving that considers problems in their entirety. This approach is built on the premise that the set of dynamically interrelating subsystems creates a synergy effect that do not exist if we consider each subsystem separately (see [RUB 01]). The same conceptual foundation has been adopted by other similar studies. Lee and Choi [LEE 03] identify the approach as suitably able to capture the complex and dynamic relationships associated with KM processes as it encapsulates problems in their entirety from a holistic perspective.

Based on a review of the literature, we present an integrated analytical model enabling research of the relationships between these factors. This model facilitates a comprehensive approach that considers the necessary processes and tools for KM and the additional factors that impact KM activities, such as leadership, technology and culture. In the theoretical framework called *ClicK.Innov* – context (leadership, information technologies (ITs) and culture), knowledge cycle and innovation performance – we assume that KM practices are driven by enabling factors that mutually interact over the knowledge lifecycle within the framework of producing new or significantly improved products. A visual representation of the model is given in Figure 6.1.

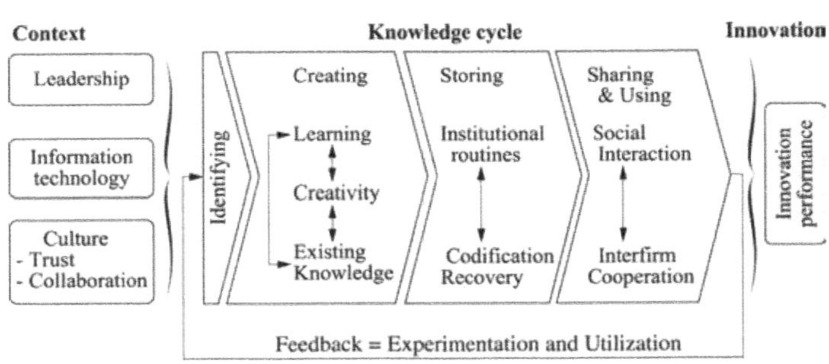

Figure 6.1. *ClikC.Innov model (Source: Santos and Wane, 2011)*

Below, we present each of the core factors included in the proposed framework, illustrating the theoretical relationships between them.

6.3. KM practices fostering innovation: what practices?

Managing knowledge requires the use of appropriate tools, processes and practices related to each other [ALA 01]. The set of activities making up the KM process is commonly known as the "knowledge lifecycle" (e.g. [MCE 00, BIR 02]). Various authors have dedicated studies to the respective phases of the KM lifecycle (e.g. [ROL 03, CEN 04]). In general, the differences between authors reside both in the identification and the number of activities and in the importance and level of detail attributed to each factor [CEN 04a]. Two fundamental stages to the KM cycle are identified: the creation of knowledge (also termed "production" or "capturing") and the sharing of knowledge (also termed "transfer" or "diffusion") (e.g. [NON 94, MCE 00, ROL 03, CEN 04, DAL 05]). In addition, the literature includes references to the planning, storage [ROL 03, CEN 04a] – as an intermediate step between the creation and sharing of knowledge – utilization [CEN 04a, DAL 05] and evaluation of knowledge (e.g. [BIR 02, ROL 03, DAL 05]). In terms of the role of applying knowledge over the course of its lifecycle, there are some divergences between the positions taken by authors. While some authors opt to isolate application as a particular phase in the cycle (e.g. [CEN 04a]), others deem this to be a transversal facet of the entire process (e.g. [MCE 01, ROL 03]). Detailed analysis of the various approaches encountered across the literature allows us to conclude that, despite the previously mention divergences, there is a great deal of complementariness between the perspectives proposed. Table 6.1 shows the features of the various approaches and then integrates them.

KM practices consist of those organizational initiatives and activities designed to manage the knowledge held by the organization's members over the course of the main stages of the knowledge lifecycle [CEN 04a, ZAC 09]. The challenge lies in understanding, for each phase, precisely which tools and techniques are most suitable to leveraging value from knowledge [BIR 02]. To identify which KM practices could be used to help organizations to achieve better innovation performance levels, it is important to look at existing empirical studies on the link between KM and organizational performance-related outcomes, although it may also be useful to establish a parallel between these practices and the knowledge cycle, so as to ensure that organizational initiatives comprise all components making up KM. We argue that an effective KM practice oriented to innovation requires a balanced implementation of activities associated with different KM processes, which is in-line with CEN [CEN 04a] and Darroch [DAR 05]. In other words, a successful implementation of KM practices fostering innovative performances should not focus only on one or two knowledge processes in isolation. For this reason, we conceptualize KM practices along four dimensions:

– Knowledge identification practices concentrate on planning KM and determining current knowledge requirements [ROL 03, CEN 04].

– Knowledge creation practices designed to boost the level of knowledge available to the organization [DAL 05].

– Storage practices related to that done to integrate (internal and external) knowledge thereby avoiding its "loss" [ROL 03].

– Knowledge sharing and utilization practices seeking to transfer knowledge to the right place, at the right time and with the appropriate quality [CEN 04a].

McElroy [MCE 00]	Birkinshaw and Sheehan [BIR 02]	Rollett [ROL 03]	CEN [CEN 04a]	Dalkir [DAL 05]	**Integrated cycle**
–	–	Plan	Identify	–	**Identify**
Production (individual / group learning) Production (validation)	Create –	Create Integrate	Create	Create & Capture	**Create**
Production (acquiring information)	Mobilize	Organize	Store	Create & Capture (contextualize)	**Store**
Knowledge validation Integration –	Diffuse – Commoditize	Transfer Maintain Assess –	Share Use –	Share & Disseminate (Assess) Acquire & Apply Update	**Share & Use**

Table 6.1. *KM cycle*

On the whole, a total of 21 KM practices were identified and grouped according to the respective lifecycle phase (see Table 6.2). In the following, we provide a brief description of each set of practices making up the integrated cycle as well as their link with innovation performance.

6.3.1. *Identification practices*

Identifying knowledge represents a decidedly strategic phase and seeks above all to align the KM strategy with the objectives and strategic targets of the organization [ROL 03, CEN 04]. The identification of knowledge involves ascertaining only which core fields of knowledge enable the organization to advance its strategic objectives in the present and into the future [CEN 04a]. KM success is only ever attainable when the organization's prevailing needs are subject to constant review and adaptation. Additionally, the accomplishment of identification knowledge practices brings about the clarification of expectations held by the parties involved and opens the way to building consensuses and commitments [ROL 03].

KM processes	KM practices
Knowledge identification	Defining and communicating KM strategy Identifying future knowledge needs Identifying current knowledge needs Implementing knowledge audits Following (and adapt) KM objectives
Knowledge creation	Using internal discussion forums Knowing the best practices performed in the sector Disseminating customer feedback within the organization Implementing brainstorming sessions Using creative problem-solving groups Capturing knowledge through seminars or conferences
Knowledge storage	Using FAQ systems Using "yellow pages" for expert location Implementing intranets or portals Using IT to forecast and simulate events
Knowledge sharing and using	Providing time to try (experiment) new methods or tools Implementing mentoring or coaching programs Involving employees in internal and/or external communities of practice (CoPs) Implementing partnerships with other companies Establishing external interactions throughout product (or service) development Using after action reviews and embedding "lessons learned"

Table 6.2. *KM practices*

Aspects deemed critical to this phase are the identification of knowledge needs in accordance with the strategic positioning [CEN 04a] and the prevailing defined general and specific objectives for the remaining phases [ROL 03]. This prospective analytical process is fundamental for justifying and supporting decision-making (e.g. on staffing needs, procuring support tool) [CEN 04a] and for defining specific KM-related objectives [ROL 03].

The KM practice of seeking to extract strategic value from knowledge is among the lesser mentioned by empirical studies. However, Zack and his colleagues [ZAC 09] found that companies implementing KM practices such as identifying knowledge sources, strategic planning (mapping techniques) and external benchmarking do tend to return better results in terms of product leadership.

This strategic planning technique is often implemented through forms of KM strategic design processes (e.g. [FOR 02]) or planning techniques (e.g. [ZAC 09]). In particular, to identify knowledge gaps, the organization may make recourse to knowledge audits, mapping techniques, surveying the specific needs of users and/or "lessons learned" [CEN 04a, CEN 04c]. The most commonly applied methods for identifying strategic knowledge gaps include techniques such as the SWOT analytical approach and feedback from clients, partners and suppliers [CEN 04a, CEN 04c].

6.3.2. *Creation practices*

The link between knowledge creation and innovation is widely supported by conceptual and empirical literature. For example, according to the SECI theory on knowledge creation, subsequently adopted in many empirical studies (e.g. [LEE 03]), the sources of innovation multiply when individuals share their tacit knowledge that is then rendered explicit [NON 94]. Seemingly, the literature allows us to claim that new knowledge creation, in conjunction with the capacity to produce new knowledge based on that existing knowledge, positively contributes toward innovative activity. The integration of existing knowledge provides for the rediscovery of internally produced knowledge, while more knowledge may be absorbed from external sources [ROL 03]. To summarize, we can say that knowledge creation practices should be focused on two critical factors: learning and creativity/fostering new ideas (e.g. [NON 94, ALL 03, ROL 03, CEN 04]).

Learning deals with the capacity to absorb the knowledge of other persons and transform it into new useful knowledge. Based upon individual learning and experiences, groups discuss ideas and create a constructive dialogue around a specific theme and thereby build up new knowledge [MCE 00]. Indeed, learning and prior related knowledge enhances the ability to recognize the value of new information, assimilate it, and apply it to commercial ends [COH 90]. The production of knowledge results from social interaction and organizations should correspondingly use tools to foster such engagement. Hence, much importance is attributed to establishing learning networks in organizations [MCA 00]. The literature also refers to other practices and tools that enhance organizational learning such as training programs, learning by doing (e.g. [CEN 04a]), discussion forums, communities of practice (CoPs) [MCE 00, BIR 02], academic journals, conferences and after action reviews/"lessons learned" (e.g. [CAL 04, CEN 04c]).

Promoting creativity and experimentation is also needed whenever we design KM practices targeting innovation, otherwise it is hard to trigger the new ideas necessary for innovation to occur. This is compatible with the view of innovation as the successful implementation of creative ideas (see [HEN 11]). In-line with this line of thinking, organizational creativity represents the means of transforming knowledge into value [LEE 03] through the development of ideas susceptible to being turned into saleable products and services [BIR 02]. Stimuli for generating new ideas derive both from internal (e.g. contacts with experts in a specific domain) and external sources (e.g. contacts with clients and other partners). Creativity may be boosted through specific tools, including: tools such as brainstorming, group problem solving, external benchmarks and CoPs, among others (e.g. [BIR 02, ROL 03, CEN 04]).

6.3.3. *Storage practices*

When new knowledge is evaluated as valid and appropriate to the organization, then it requires internal dissemination (e.g. [MCE 00, CEN 04]) as a means of extracting value from that which has been produced [BIR 02]. Even if knowledge can be stored and institutionalized through organizational routines, for now we will only look at those activities associated with knowledge codification and retrieval. The storage stage involves the selection, organization, categorization and updating of information and knowledge [CEN 04a]. In the knowledge storing process, competitive advantage derives from the effectiveness of means of accessing the information as well as the quality of the information given that such dimensions are difficult to copy [BIR 02].

The dissemination of knowledge can speed up innovation since people are exposed to new knowledge that interacts with the knowledge already held [DAR 05]. Consistent with this, some authors posit that converting to tacit to explicit knowledge is an important driver for innovation to occur (e.g. [MCA 00]). There is a corresponding need to store such knowledge in a way able to facilitate and foster searches for knowledge/learning as well as for problem solving (e.g. [ROL 03]). The critical factor of success consists of identifying the appropriate means for the codification and recovery of knowledge, without losing any of its original value [DAV 98, BIR 02].

The storage practices analyzed in the literature featured manuals, written documents (e.g. [CHO 03]) and technological tools (e.g. [MAS 02, CEN 04]). Methods and tools that support these practices include knowledge charts (displaying the knowledge held by the organization and by individuals), knowledge taxonomies, interviews with experts [DAL 05], keyword search motors [ROL 03], intranets, corporate sites, enterprise or knowledge portals and "ask–answer" (FAQ) systems [CEN 04a, CEN 04c], among others.

6.3.4. *Sharing and using practices*

On having created and stored knowledge, organizations should concentrate on internally conveying and subsequently utilizing the learning and experience. For many authors, KM adds value to an organization through optimizing the use (application) of knowledge. The main challenge facing the organization involves incorporating the knowledge produced into organizational practices inherently requiring the experimentation and sharing of knowledge between people (e.g. [BIR 02, CEN 04]). There is considerable empirical evidence for the point that the sharing and using of knowledge are crucial to leveraging the introduction of NP (e.g. [MAS 02, LEE 03, JIA 08]).

If knowledge is socially constructed, its sharing (and use) involves a large component of social and direct interaction. Similarly, given the openness of organizational boundaries due to the emergence of open innovation models, it also applies to external knowledge held by partners, suppliers or clients via external cooperation. These two facets (i.e. social interaction and external cooperation) emerge from the literature as critical variables for the sharing and use of knowledge. Therefore, among the most researched issues regarding the relationship between KM and innovation capacity are those practices related to teamwork and the use of networks to facilitate innovation [DAR 05].

Interactions between persons open up opportunities to experiment with knowledge (e.g. [BIR 02, ROL 03]) and convey the know-how existing internally (e.g. [CEN 04a]). Such knowledge sharing may occur within the framework of programs involving direct interaction, such as coaching and mentoring [CEN 04a]. However, there are other tools for nurturing such social interaction, including videoconferences, portals, discussion forums, user communities, CoPs, seminars, job rotating and contacts with experts (e.g. "centers of excellence") [BIR 02, CEN 04, CEN 04c].

Experimentation and exchanging know-how may also occur through recourse to external entities such as clients, suppliers and research centers (e.g. [MCA 00, BIR 02]). Cooperation with the exterior may take on various formats, such as strategic alliances and joint ventures, research consortia, taking out licenses [TID 03]. Contact with the external environment proves to be relevant to ascertaining the needs and expectation of clients, validating concepts/prototypes and running pilot projects (e.g. [CAL 02, BOO 10]).

Throughout these processes of internal and external interaction, organizational members build up their perceptions of the value and utility of the knowledge available [DAL 05]. It is this understanding of knowledge that enables the integration of new knowledge into organizational practices and/or the incorporation into products/services, thereby generating a return on the efforts and investment applied to the previous phases of KM cycle (e.g. [ROL 03, CEN 04]). Following this line of thinking, as knowledge is applied to organizational processes and/or incorporated into NP or services, the feedback generated encourages the emergence of new problems, new learning opportunities and nurturing individual inventive trends (e.g. [MCE 00, BIR 02]).

6.4. Enabling factors/organizational context

The successful implementation of KM practices depends on contexts favorable to the creation and transmission of knowledge within the scope of which all actors actively participate (e.g. [NON 94, CEN 04, DAL 05]). Knowledge is naturally rooted in human experiences and social contexts, thus an effective KM practice requires managing a set of different interfaces. There is a broad consensus around the existence of organizational factors that may act as drivers or brakes on KM practices, with a particular emphasis on organizational culture.

The enablers may be defined as the set of factors deemed critical to KM success and incorporate those organizational mechanisms and initiatives driving and encouraging the continuous development of knowledge [LEE 03]. Beyond culture, authors such as Davenport and Prusak [DAV 98], Lee and Choi [LEE 03], Allee [ALL 03] or CEN [CEN 04a] also identify other factors determining KM success: organizational strategy, organizational structure, leadership, individual capacities and ITs. The literature contains references to countless enabling factors (e.g. [FOR 02, CEN 04]), within the scope of which leadership, IT support and organizational culture are incorporated into our framework (see Figure 6.1).

There is consistent evidence pointing to the relevance of business models and organizational structure to enhance an effective KM (e.g. [MCA 00, FOR 02, ALL 03, CEN 04]). The organization's mission and strategy serve as stimuli conditioning the extent to which individuals actively share their knowledge (e.g. [CEN 04a, DAL 05]). In turn, the organizational structure should be flexible, containing low levels of hierarchy (e.g. [FOR 02, CEN 04c]) and encourage the "personification of knowledge" [MCA 00] – through the design of working roles and content and opening up opportunities to experiment with new tools (e.g. [BIR 02, CEN 04]).

Individual capacities are another of the factors flagged by the literature (e.g. [FOR 02, LEE 03]). Personal skills and competences are often underdeveloped and organizations and individuals need to be correspondingly interested in developing such capacities [CEN 04a]. In addition to being multifaceted and building up a "helicopter" vision of the organization and its respective contexts, employees are recommended to develop their communication skills, listen actively, adapt and engage in continuous learning processes, display research skills and the capacity to select and analyze information, etc. (e.g. [ALL 03, LEE 03, ROL 03, CEN 04]).

As discussed previously, leadership, IT and culture are incorporated into our framework. In the following, we look in more detail at each of these KM enablers.

6.4.1. *Leadership*

Leadership refers to the ability to influence a group toward the achievement of KM strategic objectives (e.g. [NON 94, CEN 04d]). According to the cited authors, leadership is understood in a broad sense, that is, it relates to top and middle management commitment and support in implementing KM initiatives.

Leadership and management support should nurture a shared vision of knowledge and enhance its importance (e.g. [NON 94]). Managers and team leaders play a critical role in KM, due to its influence on promoting new ideas, supply advice and mentoring to their teams (e.g. [NON 94, FOR 02]) and motivate their staff to actively participate in KM processes (e.g. [CEN 04c]). Leaders also encourage KM implementation by leading change processes (e.g. [FOR 02]) and establishing autonomous working teams (e.g. [NON 94, MCA 00]). These autonomous working teams are critical to KM initiatives given its role in fostering individual responsibility (e.g. [FOR 02]), and in guiding action and deliberation [NON 94] alongside employee recognition [CEN 04b].

Thereby, existing research suggests that management support is crucial for the success of KM practices (e.g. [DE 00, SVE 07]). To summarize, we can say that the organization's governance ought to be based on democratic leadership (e.g. [ALL 03]), foster the creation of autonomous work teams (e.g. [FOR 02]) and motivate members of staff toward generating new ideas and sharing their respective knowledge and learnings (e.g. [CEN 04d]).

6.4.2. *IT support*

An equally important dimension is the technological infrastructure and its ability to serve as a tool for supporting the creation and sharing of knowledge [MAS 02, CHO 03, DAL 05]. KM implementation may be backed up by technologies (e.g. [CEN 04a]) as the infrastructural solutions necessary to bringing about not only the sharing of knowledge but also providing the stimuli required to ensure that all individuals become active participants in KM processes [DAL 05].

IT is referred to throughout the literature as a tool for supporting working in groups – communication and cooperation – and for seeking out knowledge (e.g. [BIR 02, LEE 03, CEN 04]). The ease of access to knowledge and knowledge systematic storage allows knowledge reusing, thus facilitating the implementation of KM processes (e.g. [LEI 01, DU 07]). Empirical studies have also shown that KM practices focused on people and based upon IT should be carefully balanced to generate high performance levels (e.g. [CHO 03]). Nevertheless, the literature also finds that such IT infrastructures only contribute toward practices involving the codification and/or storage of knowledge (e.g. [LEE 03]).

6.4.3. *Culture*

An organizational culture based on values such as transparency, trust, collaboration and openness to new ideas proves a central facet in many studies (e.g. [DAV 98, ALL 03, LEE 03, ROL 03, CEN 04c]). Organizational culture is the greatest determinant to KM success for two reasons (e.g. [DAV 98, ALL 03, ROL 03, CEN 04b]): first, given its role in defining just what is perceived as internally valid knowledge [LEE 03] and second due to the influence wielded over other components, such as the organizational structure and the physical design of places of work [FOR 02]. As such, culture should ideally favor organizational creativity and the tolerance of errors, promoting cooperation and ensuring all knowledge is equally valued (e.g. [DAV 98, ROL 03]). Correspondingly, core factors include the levels of trust and ongoing collaboration in effect at the company.

Trust is built on maintaining mutual levels of conviction and commitment, reciprocity and the appropriate intentions, behaviors and competences for successful KM implementation [LEE 03]. Trust relationships exist when individuals reciprocally believe in the behaviors, attitudes and skills of others (e.g. [CEN 04c, SVE 02]). The lack of trust has been suggested in several studies as one of the main barriers to creating and sharing knowledge (e.g. [DAV 98, DE 00]). Researchers have also reported that high levels of trust and credibility among people contribute to accelerate knowledge flows within firms (e.g. [DE 00, CEN 04b]).

Collaboration refers to the level of mutual support, help and cooperation present in the organization [LEE 03]. Prior research has acknowledged the critical importance of collaboration in promoting knowledge creation and sharing (e.g. [NON 94, CEN 04c]). The presence of collaborative environments may increase knowledge exchange as it facilitates experimentation, helps to create a shared understanding about an organization and reduces functional silos (e.g. [FAH 98, DE 00, CEN 04d, SVE 07]). Similarly, collaboration was seen as important to enhance innovation performance through knowledge sharing and dissemination (e.g. in the course of learning networks) [MCA 00].

6.5. Innovation performance

Innovation performance seeks to measure the results of innovation and thus the actual impact of innovation-related activities on the economic and financial performance of the organization [COT 10]. The literature review ensures we may propose that there is a positive relationship between KM and innovation, although little is known about whether a combination of various KM practices generates interactive effects that enhance and boost innovation performance. Perhaps due to the lack of integrated models to implement KM, there are no stronger empirical grounds for affirming that the implementation of KM practices, mutually interlinked, leads to higher innovation performance levels than implementing them in isolation.

As indicated, the innovation performance concept is bound up with the measurement of the results arising from innovation. Indeed, the level of success achieved may be grasped through different factors (e.g. [ORG 05, MAR 11]): (1) innovation inputs (e.g. innovation objectives, barriers to innovation, the IC involved), (2) innovation activities (e.g. research and development (R&D), equipment procurement, training), (3) innovation outputs (e.g. patents, NP, processes, business models) and/or (4) innovation impacts (e.g. turnover, competitiveness, productivity). Yet, innovation inputs and activities are excluded from the scope of the proposed framework.

From the academic point of view, various authors have taken up the theme of corporate innovation capacity evaluation (e.g. [HAG 03, COT 07, JIA 08, MAR 11]), even while there is no universally accepted means of measuring the results of introducing invention into the marketplace. The literature details three practices utilized for such purposes (see Table 6.3): (1) establishing indicators for measuring sections of innovation processes (e.g. [MAS 02, HAG 03, CAL 04]), (2) specific tools integrating a set of specific innovation measurement indicators (e.g. [GPE 10, MAR 11]) and (3) IC methodologies designed to measure intangible assets (e.g. [BON 01, SVE 01, CEN 04d]). The relationship between innovation and IC deserves to be highlighted, as many authors, such as Edvinsson and Sullivan [EDV 96], considered innovation to represent a facet of the company's own structural capital, as well as KM to be integrated in human capital. Both the specific tools for measuring innovation capacities and the IC measurements are inspired on scorecard models and incorporating a range of indicators grouped into different financial and/or non-financial perspectives. The strategic nature of these tools results in the choice of the respective indicators adopted depending primarily on the strategy and the context of each organization.

<table>
<tr><th>Measuring innovation</th><th>Related literature</th></tr>
<tr><td colspan="2">(1) Measuring sections of innovation processes</td></tr>
<tr><td>– Examples of input metrics:
Top management allocation to innovation activities (percentage of time) | Sessions to stimulate creativity (no.) | R&D investments | R&D training expenditure (percentage of turnover) | Publications subscription (no.)</td><td rowspan="4">[MAR 11, BCG 10, COT 10, ORG 05, HAG 03]</td></tr>
<tr><td>– Examples of activity metrics:
R&D projects (no.) | Co-operation partnerships (no.) | Best practices workshops (no.) | R&D expenditure (percentage of turnover) | R&D personnels (FTE) | Allocation to experiment novelties (percentage of time) | Non-R&D innovation expenditures (percentage of turnover)</td></tr>
<tr><td>– Examples of output metrics:
Patents, trademarks, copyrights or designs registered (no.) | Published articles (no.) | Citations (no.) | New products (NP) (no.)</td></tr>
<tr><td>– Examples of impact metrics:
Turnover form NP and patents | Time to market | Sales of NP (percentage of turnover) | New clients/markets (no.) | ROI | Employee satisfaction | Customer satisfaction</td></tr>
<tr><td colspan="2">(2) Specific tools for innovation measurement</td></tr>
<tr><td>Community innovation survey (CIS)
The CIS collects information on European company's innovations and innovation activities. The framework for measuring innovation is based on four pillars: (1) innovation activities (product, process, marketing and organizational); (2) cooperation with other companies and research institutions; (3) legal and institutional environment and (4) interaction with the market.</td><td>[GPE 10, [ORG 05]</td></tr>
<tr><td>Innovating score
This tool enables organizations to diagnose measure and question their performance and potential of innovation. The analysis of innovation includes four main areas – conditions, resources, processes and outcomes – to which are assigned different weights resulting in an innovation score.</td><td>[COT 07]</td></tr>
<tr><td colspan="2">(3) IC methodologies</td></tr>
<tr><td>Intangible asset monitor [SVE 97]
This conceptual model comprises a large number of indicators to measure intangibles (with the exception of financial indicators). Intangible assets are grouped into three families: external structure, internal structure and individual skills. The choice of indicators depends on the organizational strategy.</td><td>[SVE 01, BON 01]</td></tr>
<tr><td>Índice CI [BON 99]
The authors proposed a combination of different indicators into a single index. This index should include two sets of measures: one set with indicators related to value creation drivers and the other comprising performance measures. The selected indicators must be aligned with organizational strategy and priorities.</td><td>[SVE 01, BON 01]</td></tr>
</table>

Table 6.3. *Measuring innovation*

In practice, measuring innovation performance levels proves to be not only a complex exercise but also difficult to implement. Innovation is a continuous and dynamic process and is consequently more difficult to measure than static activities with small and incremental innovative changes, and especially hard to capture [ORG 05]. The annual BCG questionnaire finds only 41% of executives express satisfaction about the practice implemented at their companies for measuring the results of innovation. The most commonly adopted indicators are client satisfaction and growth in profits. However, according to the cited authors, these measures are not perfect for established innovative performance since both the aforementioned metrics are influenced by a series of factors external to the organization's own innovative capacities [BOS 10].

Empirical evidence also supports that measuring performance proves to be a fairly complex exercise that does not attract much consensus. The studies analyzed include both qualitative and quantitative metrics. While qualitative measurements make reference to indicators such as employee satisfaction and creativity (e.g. [MCA 00, FOR 02]), the quantitative approaches feature both financial indicators such as business turnover and return on investment (ROI) (e.g. [ZAC 09]) and non-financial measurements such as the number of NP (e.g. [MAS 02]). All the studies analyzed resorted to the first of the aforementioned typologies (i.e. using indicators to measure parts of the innovation process) with the literature detailing innovation impact indicators (e.g. [CHO 03, CAL 04]) and combinations of indicators on activities, outputs and impacts (e.g. [MAS 02, JIA 08]).

6.6. Conclusion

In highly globalized and knowledge-based environments, with intangible assets undergoing constant change, knowledge and innovation specifically become the core sources of competitive advantage and value creation. The literature dedicated to researching the relationship between KM and product innovation highlights the existence of three core concepts: enabling factors, KM practices and innovation performance. Clarifying the relationship between these factors would seem a crucial step toward understanding only which aspects prove effective in contributing toward sustained organizational success.

From a standpoint of a broad research field, ranging from strategic management to IC, it became clear that a firm's innovative capacities to a great extent depend on its effectiveness in dealing with knowledge, as well as its ability to foster knowledge-friendly environments. It is not quite clear how the three core factors link up. In the literature, many valuable findings concerning these factors are offered, but if we see them in isolation no common theory will emerge. The problem of seeing the core factors in isolation is that many critical issues may be misunderstood, covered up or ignored. In this chapter, we highlighted the driving forces behind KM and innovation, the importance not only of KM processes but also of KM tools and practices and the KM view as a system. These issues were embodied in the *ClicK.Innov* framework that was designed to support the implementation of knowledge processes oriented to innovation, highlighting the key elements that should be included in it. Overall, the framework suggests that KM practices draw on the energy of the enablers and interact with them to generate new or significantly improved products. In this chapter, we argued that our framework developed the theoretical bases for understanding the specific relationships among the different core factors referred in the literature. It was also argued that the proposed model ensures that organizations are able to fully grasp the factors that are critical to effective KM. First, we need to perceive whether the prevailing organizational culture, managerial and technological support levels are factors encouraging KM implementation. Second, we should take into consideration that different practices use different tools. Third, organizations may carry out self-diagnosis procedures that understand whether or not the most appropriate mechanisms and initiatives are in place to maximize innovation and direct investment in those practices able to return the best results in terms of new or improved products. Finally, the performance indicators applied may result in guidance for organizations intending to implement (or improve) output and innovation impact measurement systems.

Another important feature of *Cilck.Innov* is that it draws our attention to analyze (and implement) knowledge practices comprising all components of KM, that is identification, creation, storage and sharing and using knowledge. As discussed previously, not all the stages in the KM lifecycle have been attributed due attention by the field that tends to broadly study the creation and sharing of knowledge. There is a clear need to reach out to the other phases inherent to the KM cycle so as to understand their impacts on innovation performances. For instance, the practices designed to identify knowledge are among the lesser mentioned by empirical studies. While not encountering strong empirical evidence to the identification of knowledge impacting innovation performance, the literature does concur that this phase endows organizational direction through the implementation and achievement of strategic targets and thereby contributes to the successful implementation of the other knowledge processes. Once again, it is not clear whether a combination of various KM practices generates interactive effects that enhance and boost innovation performance. Consistent with other authors, it was also argued that a successful implementation of KM practices fostering innovative performances should not focus only on one or two knowledge processes in isolation. Hence, the proposed model supplies a broad-reaching perspective of the KM cycle and enables the verification of the direct effects of the practices designed to identify, create, store and share and utilize knowledge for the purpose of innovation.

Finally, the proposed framework may provide some guidance for organizations intending to implement KM processes from a broader and more holistic perspective. This chapter contains many suggestions regarding how to foster knowledge-friendly environments and how to implement KM practices to leverage innovation product capacities. Most of these could not have been made in any systematic way without an integrative approach.

6.7. Bibliography

[ALA 01] ALAVI M., LEIDNER D., "Review: knowledge management and knowledge management systems: conceptual foundations and research issues", *MIS Quartely*, vol. 25, no. 1, pp.107–136, 2001.

[ALL 03] ALLEE V., *The Future of Knowledge: Increasing Prosperity Through Value Networks*, Butterworth-Heinemann, Amsterdam, 2003.

[BIR 02] BIRKINSHAW J., SHEEHAN T., "Managing the knowledge life cycle", *MIT Sloan Management Review*, vol. 44, no. 1, 2002.

[BOG 07] BOGNER W., BANSAL P., "Knowledge management as the basis of sustained high performance", *Journal of Management Studies*, vol. 44, no.1, pp. 165–188, 2007.

[BON 01] BONTIS N., "Assessing knowledge assets: a review of the models used to measure intellectual capital", *International Journal of Management Reviews*, vol. 3, no. 1, pp. 41–60, 2001.

[BOO 10] BOOZ & COMPANY, The 2010 innovation 1000: how the top innovators keep winning, 2010. Available at www.booz.com/global/home/what_we_think/featured_content/innovation_1000.

[BOS 10] BOSTON CONSULTING GROUP (BCG), Innovation 2010: a return to prominence and the emergence of a new world order, 2010. Available at www.bcg.com/documents/file42620.pdf.

[CAL 04] CALIGHIROU Y., KASTELLI I., TSAKANIKAS A. "Internal capabilities and external knowledge sources: complements or substitutes for innovative performance?", *Technovation*, vol. 24, no. 1, pp. 29–39, 2004.

[CEN 04a] CEN WORKSHOP AGREEMENT 14924-1, European guide to good practice in knowledge management – part 1: knowledge management framework, 2004. Available at www.cen.eu/cen/Sectors/Sectors/ISSS/CEN%20Workshop%20Agreements/Pages/Knowledge%20Management.aspx.

[CEN 04b] CEN WORKSHOP AGREEMENT 14924-2, European guide to good practice in knowledge management – part 2: organizational culture, 2004. Available at www.cen.eu/cen/Sectors/Sectors/ISSS/CEN%20Workshop%20Agreements/Pages/Knowledge%20Management.aspx.

[CEN 04c] CEN WORKSHOP AGREEMENT 14924-3, European guide to good practice in knowledge management – part 3: SME implementation, 2004. Available at www.cen.eu/cen/Sectors/Sectors/ISSS/CEN%20Workshop%20Agreements/Pages/Knowledge%20Management.aspx.

[CEN 04d] CEN WORKSHOP AGREEMENT 14924-4, European guide to good practice in knowledge management – part 4: guidelines for measuring KM, 2004. Available at http://www.cen.eu/cen/Sectors/Sectors/ISSS/CEN%20Workshop%20Agreements/Pages/Knowledge%20Management.aspx.

[CHO 03] CHOI B., LEE H., "An empirical investigation of KM styles and their effect on corporate performance", *Information & Management*, vol. 40, pp. 403–417, 2003.

[COH 90] COHEN W., LEVINTHAL D., "Absorptive capacity: a new perspective on learning and innovation", *Administrative Science Quarterly*, vol. 35, no. 1, pp. 128–152, 1990.

[COT 07] COTEC PORTUGAL, *Manual de Apoio ao Preenchimento do Sistema de Innovation Scoring da COTEC*, COTEC Portugal – Associação Empresarial para a Inovação, Portugal, 2007.

[COT 10] COTEC PORTUGAL, *Guia de boas práticas de gestão da inovação*, 2nd ed., COTEC Portugal – Associação Empresarialparaa Inovação, Portugal, 2010.

[DAL 05] DALKIR K., *Knowledge Management in Theory and Practice*, Elsevier/Butterworth Heinemann, Amsterdam, 2005.

[DAR 03] DARROCH J., "Developing a measure of knowledge management behaviours and practices", *Journal of Knowledge Management*, vol. 7, no. 5, pp. 41–54, 2003.

[DAR 05] DARROCH J., "Knowledge management, innovation and firm performance", *Journal of Knowledge Management*, vol. 9, no. 3, pp.101–115, 2005.

[DAV 98] DAVENPORT T., PRUSAK L., *Working Knowledge: How Organizations Manage What They Know*, Harvard Business School Press, Boston, MA, 1998.

[DE 00] DE LONG D., FAHEY L., "Diagnosing cultural barriers to knowledge management", *Academy of Management Executive*, vol. 14, no. 4, pp. 113–127, 2000.

[DU 07] DU PLESSIS M., "The role of knowledge management in innovation", *Journal of Knowledge Management*, vol. 11, no. 4, pp. 20–29, 2007.

[EDV 96] EDVINSSON L., SULLIVAN P., "Developing a model for managing intellectual capital", *European Management Journal*, vol. 14, no. 4, pp. 356–364, 1996.

[FAH 98] FAHEY L., PRUSAK L., "The eleven deadliest sins of knowledge management", *California Management Review*, vol. 40, no. 3, pp. 265–276, 1998.

[FOR 02] FORCADELL F., GUADAMILLAS F., "A case study on the implementation of a knowledge management strategy oriented to innovation", *Knowledge and Process Management*, vol. 9, no. 3, pp. 162–71, 2002.

[GAO 08] GAO F., LI M., CLARKE S., "Knowledge, management, and knowledge management in business operations", *Journal of Knowledge Management*, vol. 12, no. 2, pp. 3–17, 2008.

[GPE 10] GPEARI, Sumários Estatísticos CIS 2008 – Inquérito Comunitário à Inovação, 2010.

[HAG 03] HAGEDOORN J., CLOODT M., "Measuring innovative performance: is there an advantage in using multiple indicators?", *Research Policy*, vol. 32, no. 8, pp. 1365–1379, 2003.

[HEN 11] HENNESSEY B., AMABILE T., "Creativity", *Annual Review of Psychology*, vol. 61, pp. 569–598, 2011.

[HIS 09] HISLOP D., *Knowledge Management in Organizations: a Critical Introduction*, Oxford University, Oxford, 2009.

[JIA 08] JIANG X., LI Y., "An empirical investigation of knowledge management and innovative performance: the case of alliances", *Research Policy*, vol. 38, no. 2, pp. 358–368, 2008.

[LEE 03] LEE H., CHOI B., "Knowledge management enablers, processes, and organizational performance: an integrative view and empirical examination", *Journal of Management Information Systems*, vol. 20, no. 1, pp. 179–228, 2003.

[MAR 11] MARTÍNEZ J., "La innovación desde el punto de vista de la administración y dirección de empresas", in ARANCEGUI (ed.), *Indicadores de innovación y benchmarking: Reflexión y propuesta para el País Vasco*, Innobasque, Zamudio, Spain, pp. 142–297, 2011.

[MAS 02] MASSEY A., MONTOYA-WEISS M., O'DRISCOLL T., "Knowledge management in pursuit of performance: insights from Nortel networks", *MIS Quarterly*, vol. 26, no. 3, pp. 269–289, 2002.

[MCA 00] MCADAM R., "Knowledge management as a catalyst for innovation within organizations: a qualitative study", *Knowledge and Process Management*, vol. 7, no. 4, pp. 233–242, 2000.

[MCE 00] MCELROY M., "The new knowledge management", *Knowledge and Innovation: Journal of the KMCI*, vol. 1, no. 1, pp. 43–67, 2000.

[NON 00] NONAKA I., TOYAMA R., NAGATA A., "A firm as a knowledge-creating entity: a new perspective on the theory of the firm", *Industrial and Corporate Change*, vol. 9, no. 1, pp. 1–20, 2000.

[NON 94] NONAKA I., "A dynamic theory of organisational knowledge creation", *Organization Science*, vol. 5, no. 1, pp. 14–37, 1994.

[ORG 05] ORGANISATION FOR ECONOMIC CO-OPERATION AND DEVELOPMENT, Oslo Manual: Guidelines for Collection and Interpreting Innovation Data, 3rd Portuguese ed., OECD, France, 2005.

[PRA 03] PRALAHAD, C., VENKATRAM, R.,"The new frontier of experience innovation", *MIT Sloan Management Review*, Summer 2003, pp. 12–18, 2003.

[ROL 03] ROLLET H., *Knowledge Management: Processes and Technologies*, Kluwer Academic Publishers, Boston, 2003.

[RUB 01] RUBENSTEIN-MONTANO B., LIEBOWITZ J., BUCHWALTER J., *et al.*, "The knowledge management methodology team: a systems thinking framework for knowledge management", *Decision Support Systems*, vol. 31, no. 1, pp. 5–16, 2001.

[SVE 01] SVEIBY K., Methods for measuring intangible assets, 2001–2007, 2001. Available at http://www.sveiby.com/articles/IntangibleMethods.htm (accessed on 1/06/2011).

[SVE 02] SVEIBY K., SIMONS R., "Collaborative climate and effectiveness of knowledge work – an empirical study", *Journal of Knowledge Management*, vol. 6, no. 5, pp. 420–433, 2002.

[SVE 97] SVEIBY K., The New Organizational Wealth: Managing and Measuring Knowledge-based Assets, Barrett-Kohler, San Francisco, 1997.

[SVE 07] SVEIBY K., "Disabling the context for knowledge work: the role of managers' behaviours", *Management Decision*, vol. 45, no.10, pp. 1636–1655, 2007.

[TID 03] TIDD J., BESSANT J., PAVITT K., *Gestão da inovação: integração das mudanças tecnológicas, de mercado e organizações*, Monitor, Lisbon, 2003

[ZAC 09] ZACK M., MCKEEN J., SINGH S., "Knowledge management and organizational performance: an exploratory analysis", *Journal of Knowledge Management*, vol. 13, no. 6, pp. 392–409, 1999.

Chapter 7

Institutional Logics Promoting and Inhibiting Innovation

7.1. Introduction

From a Schumpeterian perspective, this chapter relates socio-cognitive and institutional logics to innovation in engineering. In contrast to "big leap" innovations such as Ford's continuous assembly line and post-Fordist "small step" continuous improvement. It distinguishes operational from organizational logics, cites cases in which these have been either dynamic or inertial and also suggests implications also for operational and organizational learning. It shows how Michael Porter and leading US firms dismissed *kaizen* style continuous improvement as a production technique, whereas its success in Japan has related to implicit social and psychological contracts. It recalls the case of Mintzberg that successful innovation can come from any level within an organization provided that higher level management is open to it yet, inversely, gives examples of how it can be blocked from above because of inertial institutional logic. It contrasts this with neglected evidence that most product innovations in civilian markets in the United States, rather than only in defense, have been sponsored by the federal government, and draws some implications for innovation policies.

In an already wide-ranging literature on institutional logics, there has been little assessment to date on how these may promote or inhibit innovation

---

Chapter written by Teresa Carla TRIGO OLIVEIRA and Stuart HOLLAND.

in engineering, nor a distinction between organizational and operational logics. Yet inertial organizational logics can mean that initiatives for product innovation can be blocked from above, as in Eastman Kodak inventing the digital camera in 1975 but deciding not to market it to avoid undermining its dominance of film and in the refusal of IBM to allow its research engineers to improve the programming of its mainframe computers reflecting its presumption that it "was" the computer industry and could afford not to do so. Initiatives for process innovation can also be blocked from above, as in General Motors (GM) leaving management of what should have been a new joint venture in motor manufacturing to Toyota and thereby learning little in terms of the operational logic of *kaizen* style continuous improvement. Management at organizational levels can also fail to identify conflicting logics at operational levels, such as between the "Hamburg" and the "Toulouse plant of Airbus Industrie" in the production of its double deck A380 that led to long delays in its commercial launch and to a major financial crisis for its parent European Aeronautical Defence and Space Company (EADS) Group.

Conflicts between organizational and operational logics parallel what Barley [BAR 06] cites as the difference between what higher level managers think they know about technical work, but mostly do not, and the claim that bureaucratic and professional forms of organizing represent conflicting principles. Bureaucracy invests social control and authority in the hierarchical structure of organizations. Professionalism is vested in occupational groups [BAR 06]. Yet this distinction may be overdrawn and some alleged "bureaucrats" prove remarkably innovative, for example, when Washington officials offset inertial organizational logic in big business and synergized different disciplines, such as physics, chemistry, materials science, biology, medicine, engineering and computers, thereby creating entirely new sectors such as nanotechnology [MAZ 11].

Mintzberg [MIN 06] has illustrated that successful innovation can come from any level within an organization provided that higher level management is open to it. This has been typical of *kaizen* style continuous improvement in leading engineering companies in Japan. So far, however, there has been a tendency to treat *kaizen* as an engineering technique and little appreciation in western management literature of its relevance to principles of psychological and social contract. Faced with downsizing, outsourcing and globalization, both principles have been in question in the West with a burgeoning literature on psychological contract more notable for perceptions of its breach than observance. Yet one of the neglected features of *kaizen* is that it combines kai – meaning improvement – with zen – meaning mutual well-being or advantage. In Japan, for a century, lifetime employment for key workers embodied the implicit institutional logics of both psychological and social contracts.

This also relates to mindsets [SEN 90]. There was little disposition by the US auto majors from the 1980s to "learn up" from *kaizen* style process innovation as they lost market share to Japanese and South Korean producers, and thereafter faced near extinction. This was paralleled by Michael Porter [POR 98] claiming that American producers had *kaizen* "coming out of their ears", but it had failed them. But such failure, combined with inertial organizational logics, was in not recognizing that *kaizen* represented an institutional philosophy and culture of management–labor relations. In contrast, where such relations are based on a "drive to survive", as they initially were at Toyota and are now also for auto assembly plants in Europe, the implicit institutional logic may be of mutual advantage from process innovation and operational learning rather than only Porter's narrower concept of competitive advantage.

The main objective of this chapter, therefore, is to show from a socio-cognitive perspective that the concept of institutional logic can be enhanced by a distinction between organizational and operational logics, the degree to which these may be either innovative or inertial and how success in the latter does not simply depend on production techniques but on mindsets and attitudes. It suggests that while such logics may be tacit, their implicit logic can be identified in a manner that may enable more effective innovation on the basis of mutual advantage in management–labor relations.

In doing so, it starts by recalling Schumpeter's stress on psychology and "creative aptitude" in innovation [SCH 49], the surfacing of tacit knowledge [POL 58, POL 62, POL 68] and the extension of this by Nonaka [NON 94, NON 95] and Ichijo and Nonaka [ICH 07]. It summarizes different perceptions of the concept of institutional logic and suggests that these can be enhanced by distinguishing explicit from implicit logic, operational from organizational logic and operational from organizational learning. It shows that such logics may initially be dynamic, as in the case of Henry Ford's "big leap" to serial mass production, yet then prove inertial, whereas others such as the "small steps" of continuous improvement, as in Japanese *kaizen*, may be able to sustain process innovation indefinitely. It then illustrates the case of the Portuguese Autoeuropa plant of Volkswagen which, by awareness of mutual advantage for management and labor from process innovation and operational learning, managed to achieve higher levels of efficiency on Volkswagen's own performance criteria than any of its plants in Germany. It also draws implications from this, challenging the presumption of some "central" European governments that "peripheral" Southern Europe cannot be competitive, and outlines implications of this for innovation policies.

7.2. Innovation from Schumpeter to Nonaka

Schumpeter [SCH 49] is well known for his concepts of "creative destruction" and the "swarming" of innovations. Yet he clearly distinguished the cognitive logic of inventing from the practical logic of innovating, in the sense that thinking or finding something new is different from developing and applying it in practice. Though simple, his distinction is ignored by those citing R&D figures as if research findings were an index of innovation, not least since patents may be a means of ring fencing rather than opening up an area; although Japanese companies do register patents, they often do not register new ideas, as with Canon of "keeping secrets in the company" [PIL 04].

Centrally, Schumpeter distinguished innovating entrepreneurs from managers. According to him entrepreneurs innovate, while managers administer what innovators have achieved. He also claimed that innovation was a prerogative of leaders and presumed that many others will follow their lead, which was his key premise in his "bunching" or "swarming" of innovation in economic cycles. He further distinguished the circular flow of demand and supply for existing goods and services, as stressed by Keynes [KEY 36], from "spontaneous and discontinuous" change when innovators come "to produce other things, or the same things by a different method" [SCH 49, pp. 64–65], or what he distinguished as product or process innovation.

Schumpeter stressed that his innovator-entrepreneurs differ in three ways from the profit maximizing entrepreneur assumed by conventional microeconomic theory. First, because innovators disregard established norms of how things should be done, rather than respecting them. Second, because innovating involves a different mindset or "aptitude" from more conventional behavior. Third, that what drives innovators if not exclusively or even the prospect of profit as such, but translating their ideas from concepts into reality in a manner that can lift whole economies and societies to higher levels of well-being [SCH 49].

Schumpeter also claimed that such aptitudes and the conduct of innovator-entrepreneurs are "accessible in very unequal measure and to relatively few people" who also classically had been "outsiders" [SCH 49, p. 81]. In this regard, he was right enough for his own time, where

Edison or Bell, or Ford, could invent a new product or process, or production system, and follow it through with companies bearing their names and dominating markets. He may still also be right in the sense that key breakthroughs in innovative ideas, products or techniques still tend to depend on a few people, such as Steve Wozniak who designed the software for the Apple II computer within 3 months [WAT 06], or Steve Jobs who marketed it, or Niklas Zennström and Janus Friis of Skype.

Yet Schumpeter's "creative aptitude" can be fostered not only by "outsiders", but also for and by insiders within the companies. As Nonaka puts it:

> "New knowledge always starts with the individual. A brilliant researcher has an insight that leads to a new patent. A middle manager's intuitive sense of market trends becomes the catalyst for an important new product concept. A shop floor worker draws on years of experience to come up with a new process innovation". [NON 94, p. 26]

Nonaka also advanced Polanyi's [POL 52, POL 62, POL 68] claims that tacit knowledge can be surfaced in the SECI model known as such from the initial letters of four stages of interfacing tacit and explicit knowledge: (1) socialization, either in informal sharing of experience or, doing so more formally, in a manner similar to Argyris and Schön's [ARG 74, ARG 78, ARG 96] reflective practice; (2) externalizing an idea or practice that may not earlier have been "out in the open" or "known by name" and thereby communicating it; (3) combining tacit and explicit knowledge, such as in developing an idea or building a prototype, which was what Henry Ford did with a small inner circle in developing his "Model T" [LAC 87]; and (4) internalization of new explicit knowledge as this becomes a part of a group knowledge base on the sense of something now "known already" and tacitly accepted as common practice.

As Thomas Kuhn [KUH 62] has stressed, intuition, youth and being "new to the field" can play a key role in an innovative paradigm shift, as it has done from the now global springboard of the Web. This has been typical of some of the fast forward innovations such as Skype, YouTube, eBay, Google, Yahoo and Facebook. Within 2 years of its launch, Skype had so challenged telecom giants, and attracted such advertising, that it had been worth $2.6 billion to Niklas Zennström and Janus Friis when they sold it to eBay. Within 18 months of its launch in February 2005, YouTube had a 100 million people daily online. This is fast-forward innovation, and dramatic. Record store sales declined by 76% between 2000 and 2010 and record stores themselves by 77% [IZZ 11, PAN 12].

Some conceptual innovations have been so fast that it is less than clear who first came up with them. In 2003, Mark Zuckerberg, who had registered at Harvard to study psychology, helped with a social Website being developed by some fellow students. After a few months he came up with the Facebook (later dropping the definitive article). Within a few years it had 400 million users. There was a challenge whether the insight giving rise to this was his or that of the student group with claims also of breach of privacy [CAR 10]. There then was again a challenge on the handling of its mega flotation as too fast and inadequately prepared [GAR 12].

7.3. Institutional logics

There is a wide range of literature with a host of claims for institutional logic, most of which vary in terms of meaning, use and context. Much of this initially stemmed from wider sociological theory such as that of Talcott Parsons [PAR 51, PAR 56] who had been influenced by Max Weber [WEB 04], Pareto [PAR 09] and Durkheim [DUR 52]. Thus Friedland and Alford [FRI 91] have seen different institutions, such as the capitalist market, governments and the State, democratic institutions, the family and religion with their own logics. Thornton and Ocasio have defined institutional logics as "the socially constructed practices, assumptions, values, beliefs and rules by which individuals produce and organize time and space, and provide meaning to their social reality" [THO 99, p. 804]. They also allow for analysis of both explicit and implicit social processes [THO 99, THO 08].

Studies of institutional logics have also included the relation of Adam Smith's concept of trust in his *Theory of Moral Sentiments* [SMI 59] to early thrift institutions [HAV 97], the transition to market logics in health-care organizations [SCO 00], similar pressures for market logics in colleges and universities [GUM 00], institutional logics in consumer research [MOO 02], mutual funds [LOU 02, LOU 07], equity markets [ZAJ 04], accounting, architecture and publishing [THO 99, THO 05] and occupational prestige [ZHO 05]. Yet some inertial institutional logics may mean that change is adopted in rhetoric rather than in practice. Lok [LOK 10] found this to be the case in the context of a British company law reform bill, enacted in 2006, which aimed to distinguish shareholder value maximization from "enlightened" shareholder value in terms of respecting employee well-being, the environment and relations with suppliers and customers. He found that both management and institutional shareholders such as pension funds adopted the rhetoric of "enlightened' shareholder value, but did not change how they acted. Other institutional logics may also prove to be negative, such as those of alleged higher professionalism in mutual funds [LOU 02, LOU 07], which replaced sensing of risk and intuition by mathematical modeling and thereby, on narrowed parameters, paved the path to the subprime crisis [MAN 06, ROU 07].

7.3.1. *Bridging, spanning and synergies*

Inversely, bridging and spanning of what are otherwise both intellectual and institutional boundaries can enable synergies that are mutually beneficial in facilitating innovation [SCO 81, BLA 04, CRO 04, NIC 11]. Swan *et al.* [SWA 02, SWA 05, SWA 07] nonetheless have recognized that success in this requires reconciliation of different interests if synergies are to be gained. This confirms Guston's [GUS 01] emphasis on the need for dialog that can surface and demonstrate mutual benefit since if boundaries are changed or shifted, some actors' interests will be reduced or displaced. Yet also that boundary spanning is feasible. Reay and Hinings [REA 09] have proposed that cognitive flexibility and integrative bargaining between top and middle management teams may strengthen the identities of key actors and allow reconciliation of competing logics. Mørk *et al.* [MØR 12], in a case study research in a Norwegian hospital, found that boundaries can be spanned across the domains of medical practices, scientific practices, industrial practices and policy practices, reconciling diverse interests in a synergic and innovative manner.

Moreover, institutional logic may happen within the institution of government itself and how it may be able to achieve "boundary spanning" between different institutions, including its own different departments and their research and innovation programs. This has been evidenced by Mazzacuto [MAZ 11], Block [BLO 08], Block and Keller [BLO 11], Lazonick [LAZ 10] and Lazonick and Tulum [LAZ 11] . Their findings show how the US federal government from the 1980s diversified funding for innovation from defense and defense-related research to civil economy sectors and created entirely new sectors such as nanotechnology and "orphan drugs". Major pharmaceutical companies would not pioneer since, with a potential market of only 200,000 or fewer sales, they did not deem it worth their while, whereas sponsoring them then generated related drugs in the sense of Schumpeter's swarming and clustering, with mass market reach.

What this suggests is that the federal government offsets inertial institutional logic in the corporate sector since big business could not, or would not, manage "fleet-foot" innovation because of what Deeg [DEE 05] has deemed "path dependence" on its own research agenda. In-line with Mørk *et al.* [MØR 12], the federal government was more flexible and dynamic in understanding the connections between different disciplines – physics, chemistry, materials science, biology, medicine, engineering and computers – and their innovation potential. The under-recognized outcome was that, between 1971 and 2006, 77 out of the most important 88 innovations rated by R&D Magazine's annual awards were found to have been fully dependent on federal support, especially, and not only in promoting new high-tech start-ups, but also support and promotion thereafter [MAZ 11, BLO 11].

7.4. Socio-cognitive and institutional logics

It may also be that the conceptual framework of institutional logic can be enhanced in socio-cognitive, operational and organizational terms by distinguishing sets and subsets of meaning which are either explicit or implicit within it:

1) Conscious logic such as Taylor's [TAY 11] claims for scientific management or Sloan's adoption of the multi-divisional management model at GM.

2) Unconscious logic including Kleinian displacement and denial [KLE 32, KLE 52] which Melanie Klein developed in terms of child psychology yet which others such as Schneider [SCH 75] and Richards [RIC 89] have related to market behavior and of which examples are displacement and denial of *kaizen* style continuous improvement.

3) Explicit logic as in speech, writing, employment or other contracts, or mission statements.

4) Implicit logic in discourse [OLI 00, OLI 01, OLI 02] or in a course of action, which may or may not be recognized, such as the US auto majors relying for too long on Fordist economies of scale rather than adopting post Fordist economies of scope [OLI 07a].

5) Tacit rules in behavior which are taken for granted and assumed and which may be dysfunctional yet have never been questioned [OLI 07a].

6) Personal, normative and practical logics in the sense of what is wanted by an individual, what normally is done and how it is done [BOU 77, BOU 84, BOU 90].

7) Organizational logic that may be innovative or inertial, explicit or implicit in an organization that may be part of a larger institution such as a subsidiary or plant of a transnational corporation [OLI 07].

8) Operational logics in the sense of how things are done in normative and practical terms and may be either explicit or implicit, inertial as with single task Taylorism or innovative as in post-Fordist multitasking, multiskilling and effective group working.

7.5. Big leap innovation: Fordism

"Fordism" as a concept was devised by the Italian Marxist Antonio Gramsci [GRA 75] as combining Ford's mass production with high-income mass consumption. This was Ford's own ambition, since he was not only an innovative mechanical engineer but also a social engineer and driven by social values. He wanted a society in which not only an elite but also most people could gain from the innovation in modern engineering that was possible by the early 20th Century.

The claim that Ford was driven by values may seem misplaced for a man whose system proved so alienating for his workers that many of them quit within days, who spied on them, as parodied in Charlie Chaplin's Modern Times, who hired the local mafia to break up trade union meetings and who could not delegate. Ford's projective identification with the Model T and refusal for too long to replace it nearly broke his company. Ford was also anti-semitic to the point of being quoted with approval by Hitler in *Mein Kampf*. Nonetheless it was values that drove him from childhood, through early life, and from work experience including his aim, well before Hitler, to produce a "people's car" [LAC 87].

Yet Ford's "big leap" to a moving assembly line was not the cause of the initial success of his iconic Model T, which sold millions. Rather, its success caused him to speed up production to meet demand for its innovative design, components, performance and price. It had a light and highly flexible chassis made up of a compound of vanadium and steel, which reduced the risk of broken axles; a single cast engine block rather than separately cast cylinders which in other cars at the time tended to come apart; a magneto rather than a battery which could be started anywhere by swinging a starter handle, rather than needing recharging if it went flat, which was vital in rural areas which lacked electricity. Within weeks of its launch he had orders that would have kept his Highland Park plant near Detroit at full three shift capacity for more than a year. He knew that he had to speed up production and extended assembly line production from his magnetos to the whole production process [LAC 87].

Nor did Ford's assembly line stem from Taylorism, even if his success with it was aided by Taylor's [TAY 11] time and motion principles. It was his iterative operational logic, extended incrementally in his initial Highland Park factory and then in his Rover Rouge plant, which became the production logic of "Fordism". He then extended this to all materials and component supply through operations that he bought or built, from blast furnaces producing iron, other plants converting it into steel with the vanadium component, glass, canvas and backward linkages through to owning rubber plantations for tires. Each, wherever feasible, was on the operational logic of the assembly line principle which was his insight, not Taylor's [LAC 87], while the assembly line also influenced the engineering logic of industrial architecture with more extensive use of prestressed concrete in extended single-story factories outside urban centers becoming the new norm [BUC 02].

7.5.1. *Inertial operational and organizational logic*

Ford's organizational logic therefore was the sum of his operational logic. He needed a continuous assembly line to match high demand for his Model T. Yet a Fordist organization with a top-down Taylorist division of labor cannot readily generate organizational change. Unless it delves-down to "learn up", [OLI 07] it will not be able to draw on the tacit knowledge, skills and experience of those who may know best what is going wrong yet need not, what is wrong with how things are being done and how they could be put right to mutual advantage, nor interface tacit knowledge and explicit knowledge in the manner that Nonaka [NON 94, NON 95] and Ichijo and Nonaka [ICH 07] have stressed as vital for creativity and innovation. For when organizational logic is the sum of operational logic, and this is designed top-down on Taylorist principles, little "base up" or "lateral" learning, such as boundary spanning, is feasible. Fordist operational and organizational logics thereby are inertial. Turning Ford on this was slow, painful for those around him and by the late 1920s risked bankrupting his company [LAC 87].

7.6. Small step innovation: post-Fordism

A distinction made by Colenso [COL 00] between Fordism and post-Fordism is between the "big leap" of Ford's achievement of serial mass production and the "small steps" by which Toyota pioneered customized flexible production. Underlying this was the difference between inflexible economies of scale – doing more of the same at lower cost – and economies of scope – doing more with what we already have.

The term post-Fordism was influenced by the low volume but high-quality flexible specialization claimed by Piore and Sabel [PIO 84] in their analysis of Italian industrial districts. But it is the high volume, customized "lean production" pioneered by Toyota and analyzed in the MIT world auto study "The Machine that Changed the World" [WOM 90] that has made a major impact on management thinking.

Originally a textile machinery company, and then into cars and trucks for the military in the later 1930s and during Second World War, Toyota, in 1946, was only producing as many vehicles a year as GM was producing in a day. By 2006 it was overtaking GM as the world's no. 1 vehicles producer [NAK 06]. But this was to take time, formidable lateral thinking, a bitter industrial dispute which only then led to consensus, and a later "iterative" path, longer than Ford's mass production revolution but one which, within 30 years, was to displace it [OHN 88a, OHN 88b].

This slow start was not least because although Toyota's trucks were good enough, its cars were derived from models for the military, designed for field use, heavy to steer and also too expensive because of their low production volume than many Japanese, after the defeat of Japan, could afford [WOM 90]. It could not deliver a car that people rushed to buy as the US public had for Ford's Model T, not least since few people had even more than subsistence levels of income.

Toyota transformed this not only by making vehicles at lower cost by economies of scope rather than scale, but also by an operational logic of "fault-free" production, making them attractive because because they were reliable[1]. The achievement of fault-free assembly by Toyota stemmed in part from workers being able to stop an assembly line to prevent a fault going through, which had been anathema to Ford. It also achieved multiple economies of

1 This was challenged when there were widespread claims of an accelerator fault in one of Toyota's Lexus models, for which the company initially paid compensation. Yet a congressional enquiry found that the fault was neither in the engineering of the pedal nor assembly rather than in driver error combined with drivers wrongly using winter all weather floor mats which were not designed for and did not properly fit the Lexus model in question [CSE 11, NHT 11].

scope – doing more with the same equipment – rather than economies of scale – producing more of the same product – and thereby "lean production" [WOM 90]. An example of this was *kanban* or just-in-time delivery of components to the production line, which reduced stocks [COL 00] which we suggest relates to the distinction of operational from organizational logics. *Kanban* was first pioneered US retailers in terms of automatic replacement of stocks yet never "learned up" by US manufacturers.

Another "small step" in post-Fordist operational learning was the reduction in time it took to change a die for a multi-tonne press to shape body parts from cold steel. For the US auto majors this was during the 3 weeks summer vacation for production workers. In due course, Toyota managed to get the heavy dies for pressing cold steel changed in 3 minutes [KAP 94]. It did so by sliding them horizontally on rollers to fit under the presses, and then attaching them with heavy clips for one or more shifts, or even during a shift. What made this operational logic revolutionary for Toyota was that when more than one vehicle could be made, even in one shift, it could become a multiproduct company by more flexible use of the same steel press and assembly line.

7.6.1. *Kaizen and mutual advantage*

Behind this fault-free operational logic was the "philosophy of *kaizen*". This, nearly without exception, is just translated as "continuous improvement". Yet, as indicated at the outset, *kaizen* in Japanese combines kai as "improvement" with zen as "mutual advantage". The initial *kaizen* breakthrough was mainly multiple uses of the same tools and equipment, and cutting out wasted time and effort.

Such post-Fordist flexible production also implied interrelated social and operational logics which were crucial to its success:

1) The second principle was the founding logic, already well recognized in Japan since the second decade of the 20th Century was "lifetime employment" in the sense of assuring core workers jobs until they were 55 [YAM 04]. This was implicit, in the sense of a psychological contract, to which we return.

2) The second principle was the antithesis of Taylorism by combining multiskilling and multitasking with job rotation.

3) The third principle was the bonus system in Toyota and other leading Japanese firms which included profit sharing. This made a reality rather than

rhetoric of the concept of stakeholding. Because workers had a vested interest in the success of the firm, they also thereby achieved multiple job-fit and organization-fit rather than the alienation of Taylorist single tasking.

4) The fourth, and explicit, logic was the right of employees or work groups to be directly paid for suggestions leading to continuous improvement. But companies such as Toyota did not gain phenomenal rates of continuous improvement just because workers were paid to suggest them, not least when the rate in itself was not high [KAP 94], but because the commitment to lifetime employment meant that, in doing so, they knew they would not be innovating themselves or colleagues out of a job.

5) Within the combination of lifetime employment with profit-related pay was an implicit institutional logic of "equity" in terms of both procedural and distributive justice. A share in profits implied distributive justice in that workers knew that their contribution to a new proposal to raise efficiency meant mutual advantage both for the company and themselves.

This is not to romanticize the Toyotist model. Although workers had multiple job-fit and organization-fit, they needed to fit their lives to the needs of the organization and its operational schedules. There was next to no concern for work–life balance. Nonetheless, as Colenso stresses: "at the heart of *kaizen* lies a set of values" [COL 00, p. 158], including a mutual commitment to valuing the work of others, rather than management treating labor as only a disposable factor of production as with flexible labor market principles. As Kaplinsky and Posthuma [KAP 94] put it, the key to *kaizen*:

> "lies in the social rather than the technical domain, for the essential principles of these Japanese techniques – their technics – are easily comprehended and not technologically complex ... their successful adoption overturns many of the social relations of domination which were so important in the evolution of Taylorist forms of production management and control". [KAP 94, p. 285]

7.6.2. *Porter dismisses kaizen*

Remarkably little of the synergic operational logic and implicit equity underlying *kaizen* and continuous improvement in Japan has been recognized by Michael Porter in successive variants of his competitive advantage model [POR 80, POR 90, POR 98]. Porter asserted that *kaizen* and flexible production techniques "turned out to be illusions":

> "[C]ontinuous improvement has been etched on managers' brains. But its tools unwittingly draw companies towards imitation and homogeneity. [POR 98, pp. 42–43, 45]"

Yet if continuous improvement was "etched on managers' brains", as Porter claims, in too many cases its logic failed to penetrate them. Porter's reference to the "tools" of continuous improvement indicates that he sees *kaizen* as a management add-on rather than as a transformation of operational logic and as a means for mutual employer–employee advantage. He has also claimed that "Japanese companies rarely have strategies" rather than "imitating and emulating one another", and that:

> "Gradually, managers have let operational effectiveness supplant strategy (resulting in) … zero sum competition, static or declining prices, and pressures on costs that compromise companies' ability to invest in the business for long term". [POR 98, p. 45]

This claim displaces the ongoing long-term success of leading Japanese firms and stays within the mindset of US management theory since Chandler [CHA 62], that it is strategy that counts over all else, despite successive critiques of this over decades by others such as Mintzberg [MIN 94, MIN 04, MIN 95]. It also dismisses the degree to which the success of leading Japanese companies such as Toyota and Honda has not only been by process innovation but also long-term commitment to product innovation, as with hybrid vehicles which took more than two decades to develop and then successfully market, whereas the US auto majors, when filing for Chapter 11 protection against insolvency [SPE 12], were calling on the federal government to finance research into them.

7.7. Social and psychological contracts

Social contract theory flourished in the 17th and 18th Centuries and is most readily identified with Jean-Jacques Rousseau [ROU 62], although there had been earlier precedents for it in others such as John Locke [LOC 94]. Such theory claimed natural rights rather than the divine right of monarchs. Rousseau's case for it inspired the 1789 French Revolution [COB 98, DE 47]. This was vehemently criticized after the Second World War by Popper [POP 45] and Talmon [TAL 61] who claimed that it had been among the causes not only of the Terror in France during the French Revolution but also for the rise of 20th-Century totalitarianism.

One of the key weaknesses of such claims by Popper and Talmon is that the principles of a social contract based on natural rights, especially in Locke, directly influenced the Declaration of Independence and the Constitution of the United States, which so far has not succumbed to totalitarianism. Yet a social contract either may be explicit, as recommended by Rousseau, or implicit, as in the presumption that a government has not only a responsibility to protect property rights, as stressed by Locke, but also to assure high levels of public services and employment. Both Popper and Talmon also neglected a crucial condition in Rousseau's Contrat Social, stressed by De Jouvenal, which is more relevant to the interfacing of social and psychological contracts in that he insisted that, for a social contract to work, everyone had to know everyone else and mutually agree shared values, including the surfacing of tacit values in which he was two centuries in advance of Polanyi [POL 58, POL 62, POL 69], Nonaka [NON 94, NON 95] and Ichijo and Nonaka [ICH 07].

7.7.1. *Psychological contracts*

With the onset of globalization, and of downsizing, outsourcing and business re-engineering there has been a burgeoning literature on psychological contracts. Such a contract differs from a legal contract, or what Locke and Rousseau assumed should be the case in a social contract, in that no one writes it down rather than writes about it. It is assumed to be implicit at varying levels of consciousness. Its theory varies in what it claims, where it focuses attention and also how it relates to different realties for different people. For instance, in principle such a contract can include mutual trust, shared values and rewards, competence recognition, skill enhancement and the degree to which these may improve operational and organizational performance (e.g. [GUE 95, GUE 98a, GUE 98b, GUE 99, GUE 03, GUE 04a, GUE 04b]).

Yet psychological contracts have no formal basis either in job descriptions or employment contracts, or recourse to law, while formal contracts are getting shorter, with less commitment from employers to career development, or even any kind of career. In the new local and global climate, and since psychological contracts are not even explicit, are they really worth more than the paper they were not written on? For Rousseau [ROU 89, ROU 95, ROU 98], Rousseau and Parks [ROU 93] and Rousseau and Tinsley [ROU 97] the notable feature of a psychological contract has been individual perception of its breach.

As Guest [GUE 98b] has suggested, a one-sided contract is also inherently contradictory. If a contract, whether explicit or implicit, is concerned with reciprocal obligations, how each of the parties regard this reciprocal process are of equal relevance [GUE 98b]. It is also likely to be the case that the dynamics of interaction between the two parties will influence not only perceptions of mutuality ("it is the perception of mutuality and not necessarily mutuality in fact that gives rise to creation of a psychological contract" [ROU 95, pp. 10, 20–21]), but whether it works in practice to mutual advantage, or breaks down with not only negative consequences for an individual but for the organization as a whole.

While this debate ranges across, and in some case rages in, western management literature, there has been little reference within it to what amounts to both psychological and social contracts, in the sense of both Locke and Jean-Jacques Rousseau, from a century of management–labor relations in Japan. For example, Yamaguchi [YAM 04] and Moriguchi and Ono [MOR 04] have drawn on available studies of lifetime employment for a century in Japan and Moriguchi and Ono have described it as:

> "an economic as well as social institution, characterized by an implicit contract and reciprocal exchange of trust, goodwill, and commitment between employers and workers... (which) emerged as an equilibrium outcome of the dynamic interactions among management, labour, and government and became an integral part of the nation's employment system over the past hundred years, reinforced by complementary institutions such as state welfare policies, labour laws, corporate governance, social norms, family values, and the education system." [MOR 04, p. 1]

The text of the above paragraph from Moriguchi and Ono is not difficult to understand at face value. But, although they do not conceptualize it as such, it contains 20 "sets of meaning" implicit in the two words "lifetime employment", which concern both psychological and social contract, as illustrated in Figure 7.1.

Thus, although Moriguchi and Ono did not relate their summary of meta-analyses of lifetime employment to the concept of psychological contract, or to the concept of social contract, when deconstructing their single paragraph, there are 10 subsets of meaning within it that relate, on the left of Figure 7.1, to psychological contract, and on the right of Figure 7.1.

Psychological contract	**Social contract**
Management	Government
Corporate governance	National employment system
The firm as economic institution	Historical context
The firm as a social institution	Complementary institutions
Labor	State welfare policies
Implicit contract	Social norms
Reciprocal exchange	Family values
Trust and goodwill	The education system
Commitment	Integral employment Policy
Equilibrium	Dynamic interactions

Figure 7.1. *Psychological and social contract in lifetime employment [MOR 04]*

Neither the concept of psychological contract nor of social contract is explicit in mainstream literature on *kaizen* in Japan, such as that of Colenso [COL 00] or Womack and Jones [WOM 05]. Yet, this suggests a validation of the distinction within the general concept of institutional logics between sociocognitive logics, which may be either explicit or implicit and which, in symbiosis, have proved dynamic and innovative rather than inertial.

7.8. Inertial organizational logic

An example of inertial institutional logic is what Deeg [DEE 05] has deemed "path dependency" where a business gets "locked into" a technology strategy that it has already developed and is not open either to alternatives or even consciously blocks an innovation from within. An example of the former is the Xerox photocopying process whose inventor tried to sell it to a range of companies without success, including Kodak. An example of the latter is Kodak's R&D unit developing a digital camera in 1975 but not marketing because it could undermine its market hegemony in film [SPE 12].

In the longer term, Kodak has found its core film processing market decimated by the digital camera when this was produced and marketed by others. Kodak then was hit by a failure to compete with the rise of cameras in mobile phones by not recognizing that why people take photographs changes over time. When it introduced a $1 box camera in the 1890s, this was mainly to make a record, such as for family albums. But with the emergence of social Websites, this shifted to communication [GLA 12]. Anyone with a digital camera can select those photographs that he or she wishes to transmit either to friends or relatives via the Internet without bothering to print at all. As an outcome, Kodak's previous hegemony in film is near species extinction and, like the US auto majors, it also has been driven to file for Chapter 11 protection against insolvency [SPE 12].

Box 7.1

From Any Level: Mintzberg on Leadership and Innovation

Henry Mintzberg [MIN 06] has argued that the intention of leadership may be to empower people, but often does the reverse and thereby abets rather than aids innovative thinking. In doing so he takes the case of a report in Fortune Magazine in 1997 that in four years IBM's new chief Lou Gerstner had added more than €40 billion to the company's share value by getting into e-business, commenting that that the report implied that he did this "all by himself".

Mintzberg then cites that how IBM got into e-business was "because a programmer with an idea conveyed it to a staff manager with more insight than budget, and he stitched together a team that drove the change. And what role did Mr. Gerstner play? When he eventually heard about the initiative, he encouraged it. That's all. Instead of setting direction, he supported the direction setting of others".

Contrasting this with the heroic leadership so commonly portrayed in the press of "the great one who rides in on the white horse to save the day, changing everything at will, even if he or she only arrived yesterday, with barely any knowledge of the organization, its history or its culture".

Mintzberg claims that there is a need for more of what has been called "distributed leadership", meaning that the leadership role "is fluid, shared by various people in a group according to their capabilities as conditions change".

Asking "Is that not how the Linux Operating System and Wikipedia work?", he nonetheless suggests that even this slights what actually is happening since leadership is not "distributed" top down, and does not lie "in any individual so much as in the collective social process – essentially in community".

Source: Mintzberg [MIN 06].

IBM also is a classic case of blocking innovation from above. For years, it actually instructed that no design engineer was to seek to improve its memory base for mainframe computers, which was a case of denying improved operational logic since it would mean a whole reconfiguration of its organizational logic. This meant that it stayed locked in the mindset that computers by definition were as large as a supermarket deep freeze, needing a temperature controlled room and, since heavy, also reinforced floors to support their weight.

IBM was big, thought big and was known as "Big Blue". Yet, being locked into such a mindset of size and scale rather than scope, and the presumption that its competitive advantage would last forever, it allowed fleeter footed, highly intuitive creators in new start-ups, such as Wozniak and Jobs at Apple, to start what was to transform the whole computer market [WAT 06]. By allowing an as yet scarcely known Bill Gates to keep the rights to its software, "Big Blue" also missed what was going to add value for the new generation of computers, allowing Microsoft to overtake it. IBM also never really made a success of laptops, and in due course, gave up on personal computers, selling its business division for them to China's Lenova [MIT 07].

For years, "Big Blue" never really learned to think laterally or reach down to "learn up". Its inertial mindset and its presumption, as with the US auto majors, that it "owned" the computer industry, inhibited it from doing so. Even when innovative thinking did break through from below, as in the example drawn from IBM by Mintzberg [MIN 06] and cited in Box 7.1, it was easier to support it when it was clear that it was succeeding than to delve down and risk backing it before it did so.

7.9. Inertial operational logic

GM paradigmatically fits the case of inertial operational logic. It considered continuous improvement something that line managers should learn and left the operational management of the New United Motor Manufacturing Initiative (NUMMI) joint venture in California with Toyota entirely to Toyota.

The concern of Toyota to gain union cooperation was indicated by a comment from the environment manager of the NUMMI plant that:

> "The unions are our best allies in the plant. Many companies in the United States don't recognize that their best resources are the unions and their people. [ROT 03, p. 1789]"

It was only 25 years later that GM really began to address the issue of gaining changes in operational logic and doing so on the basis of gaining consent for it from the United Auto Workers [SIM 06b], and by then it was too late.

Even when there is "big leap" in innovation, as with Ford needing to achieve serial mass production to respond to unprecedented demand for his Model T, a "plateau effect" in both operational and organizational learning may also be obtained in when managers achieve an initial top-down change in either organizational or operational logic, then find that further top-down change-on-change becomes progressively more difficult as it encounters increased resistance from below. Or they may satisfy themselves in that they have managed big leap change and now can simply administer it. Stefan Stern [STE 06] has observed that what is implied is both organizational yet also deeply psychological. He quotes a change-management consultant on the difference between the excitement of the innovative phase of a product or service, and the next more routine phase, who observed:

> Companies seem to hit a "second brick wall"... when business just is not fun anymore. It is all administration, with little hands-on action; all rational analysis and no enjoyment. [STE 06, p. 8]

Box 7.2

Not Operational Learning: Toyota, GM and the NUMMI Initiative

– Toyota from the start, before the NUMMI plant opened, worked closely with trades unions.

– Where there was formal training it was "customised" to what the needs of the plant would be from workers whose provisional job designations and roles already were defined. Later on-the-job training was combined with a strong ethos of consultation.

– Suggestions for continuous improvement were "on line" within work teams to each other. Half of the learning gains of continuous improvement were from individuals who spent most of their time on the production line.

– Rather than only total quality control on the line, and total quality circles, work group progress was focused on problem-solving circles. Group learning in operational units was matched by lateral learning with other units within the plant.

– When productivity gains meant reduced labour input for a particular task or process, workers were redeployed elsewhere in the plant, where necessary with customised retraining, which was consistent with principles of psychological contract.

> – A conscious ethos of employee empowerment and self-direction in methods of work operation freed up specialist staff for other work such as new process design and implementation.
>
> – Toyota was so confident of success in the plant that it committed itself to a "no lay off" policy.
>
> *Source*: Rothenberg [ROT 03].

But how many managers are consciously aware of such a "plateau effect" not least since Kuhn's [KUH 62] point, matched by Senge [SEN 90], is that people may be unaware of the degree to which they think within the parameters of a given paradigm or mindset?

This will depend in part on their level within an organization. Top management itself may not be able to achieve much more than it has in an initial "big leap" change unless it can offset tacit resistance to further by gaining employee involvement and commitment to continuous improvement. It also may simply dismiss operational learning because it considers a technique needed to be learned only by online managers, as was the case in a Fordist era. In contrast, those who know most about what needs to be done are at middle or lower operational levels, yet because of an inertial organizational logic may not be able to feed this up to a higher management level.

7.10. Conflicting operational logics

A paradigmatic example of conflicting operational logics led to a major crisis for Airbus Industrie in 2006. Success had been dramatic for Airbus in challenging what had been dominance of the world civil jet aircraft market by Boeing. Compared with a global market share of less than a fifth in 1995, within a decade it had gained half and was overtaking Boeing. Its A380 double-decker jumbo had flown while Boeing's was still an idea [HOL 06].

After a successful enough demonstration flight of its double-decker A380, it had to delay its commercial launch since the components of its fly-by-wire system produced, in principle jointly, by its Hamburg and Toulouse partners, proved incompatible. Part of the problem for Airbus was a management struggle at board level in EADS, which distracted it, but not exclusively. Each success encouraged the norm of not admitting that schedules might not be met because of technical constraints [CLA 09]. As one senior Airbus executive based at the Toulouse plant said of red, yellow or green labels stuck on the information boards tracking progress on current projects: “There was a tendency for people to report ‘green’ because if they reported problems someone would come along and kick lumps out of them”. As a German trades union leader at the Airbus Hamburg plant put it:

> “You had your goals and you were meant to meet them. Any ability to notice problems was totally lacking. This was understandable in part. Every month we set a new sales record. What on earth could go wrong? [HOL 06].”

In Hamburg, teams of engineers were struggling to fit 500 km of cabling needed to connect the cabin of the aircraft with the controls for its flaps and other systems. But the tacit rule and implicit norm that deadlines should be met meant the elementary mistake of trying to do so without blueprints because there had not been time to work through design changes relative to earlier Airbus fly-by-wire models. Wiring was put in and had to be taken out again because “no one at operational level was prepared to say: ‘stop, we can’t meet our own schedule’” [HOL 06]. People unconsciously symmetrized with the presumption of success rather than “going asymmetric”, standing out and enabling “learning up”.

7.11. Operational logic and learning

One of the claims made during the Eurozone crisis was that southern Europe was irredeemably uncompetitive. Yet, this has been disproved by the case of Autoeuropa, a Volkswagen vehicle pressing and assembly plant located at Palmela on the Atlantic coast just south of Lisbon. Autoeuropa initially was the outcome of a joint venture by Volkswagen, Ford and SEAT. This decision to invest in a new green field site had been taken jointly by VW (owner of SEAT) and Ford on multiple criteria including not only the European Union approving relevant grants but also the cost of labor, the willingness to cooperate national trades unions, transport accessibility from an adjacent port and dedicated handling facilities.

One way had been to minimize costs, since real wage and benefit levels in Portugal were a third of those of Ford in the United States and Volkswagen in Germany, and notably lower than those for Volkswagen’s SEAT plant in Catalonia. Volkswagen had insisted on a German style Works Council for the plant, to which the main trades unions in Portugal agreed. Following enlargement of the EU to 25 member states, SEAT decided to open a plant in Slovakia and indicated that it would withdraw from the joint venture, as did Ford.

Our interest in and learning from the case of Autoeuropa, since early 2000, was enabled by a series of site visits and meetings with employees in the cold steel pressing plant and the various stages of the assembly process itself, including dialog with team leaders, quality circle groups, plus one-to-one meetings with operational managers including the cost controller, the production managers and employees of the pressing plant and the serial stages of vehicle assembly, as well as the purchasing and sales managers, followed by four workshops at the Faculty of Economics of the University of Coimbra, each attended by plant managers and those of its principal Portuguese suppliers. All of the workshops were audio and videotaped. They involved different managers presenting different dimensions of the operational philosophy of Volkswagen itself, and the criteria it required for performance and quality control.

As became clear from the workshops, none of the plant managers at the time of the study were familiar with concepts such as Fordism or post-Fordism, or economies of scope rather than economies of scale, yet were daily living this reality. Nor were any of them familiar with theories of tacit knowledge or implicit learning, and surfacing it to gain continuous improvement, although this is what they were doing. Nor was this was due to lack of intellect. For example, the cost controller of the plant had had a high profile career, going straight from university into the finance ministry of Portugal. But her degree had not been in management and none of the formal education by which she had gained such professional success had introduced her to any such concepts. What she and others learned was an interfacing of explicit learning from formal training and implicit learning-from-work [OLI 07].

7.11.1. *Interfacing tacit and explicit knowledge*

In-line with Nonaka's [NON 94] SECI model, some "new knowledge" in the Autoeuropa case was resurfacing earlier knowledge, such as a manager who had been a trainee in the paint workshop joining a group discussion to solve a problem in it and helping do so by a method that had been devised earlier when he/she was a trainee but which has been neglected since then. This then was assisted by cross-team meetings in which the "learning up" was both explicit and from tacit to explicit. The frontier between explicit and tacit knowledge was constantly moving, and needed capturing and recapturing to be shared. Yet, the overall outcome was lateral learning of an operational logic of flexible production in terms of multitasking and multiskilling, a commitment to continuous improvement and commitment also to employees as an asset rather than merely a cost.

As with Toyota after WW2 when it was threatened with bankruptcy by pressure from its banks [WOM 90], in the case of Autoeuropa, this was driven not by an explicit but by an implicit survival logic. Initially, and for several years thereafter, Autoeuropa was a one-car plant, at the time of the study producing only a people-carrier, the Sharan. Although its labor costs were similar to those available to Volkswagen in central European locations such as Slovakia, it needed to ensure a high level of operational innovation to avoid closure and relocation of the kind already evidenced some years earlier by Renault first coming to, and then quitting Portugal, and by a GM assembly plant whose future at the time of the study was known to be in question, and which GM then announced that it would close.

Strikingly, Autoeuropa became more efficient in terms of VW's own performance criteria than any of its plants in Germany including the prestigious Wolfsburg. It was allocated the Sirocco to replace the by then outdated Sharan, and in 2007 was scheduled to assemble also the Eos model, with employment increasing to 3,700. In the spring of 2008, it also was allocated the Golf Cabrio, a new Sharan replacement and assurance of being a four-car plant from 2010, each sharing the same basic platform. Meanwhile, SEAT had decided to relocate its Slovak operation back to higher wage cost Catalonia since the same tacit knowledge and implicit learning by Catalan workers had not been matched in Slovakia, and outstripped differentials in labor costs.

Most of the managers in the study were unaware of the principle of innovation-by-agreement, which had been endorsed in the Lisbon Agenda by a European Council only a few kilometers from their Palmela plant [EUR 00, HOL 00, OLI 07] or of the principle of work–life balance in [EUR 00], which had been recommended in the Lisbon Agenda or of the theory of "psychological contract". They had worked it out for themselves. When there was decline in demand for its people-carrier, which had reached maturity in its own product cycle, those employees not needed for production were given a combination of extended paid leave and work-sharing rather than laid off. All employees were retained on the payroll, and encouraged to commit to retraining and skill extension programs. This suited both operational and personal needs. They could combine the "free time" in terms of reduced production to enhance personal and family time, with short course group skill enhancement or another course.

This "no lay-off" commitment was less than Japanese commitment to lifetime employment, but its implicit logic was that management valued employees, rather than presuming that the labor was a variable cost that could be dismissed in times of low demand and then perhaps re-employed later, or left unemployed. Like Toyota after WW2, which for a while in the later 1940s faced bankruptcy, both plant management and other employees at Autoeuropa had a mutual advantage in survival and in gaining joint operational learning to assure this. It had done so by an implicit "psychological contract" with employees including concern with work–life balance and sustained assurance of employment not only so long as the survival of the plant itself was assured, but to assure it. None of the managers had conceptualized such a contract, rather than evolved it with the joint trades unions in its German style works council. The practical success of dialog with and through the works council was less "learned down" from VW than "learned up" on a trajectory of innovative proposals for new methods of work operation.

Box 7.3

Operational Learning in a "Peripheral" Car Plant

Volkswagen's Autoeuropa factory just south of Lisbon initially was a one-car pressing and assembly plant. The plant managers were unaware of any conceptual distinctions such as Fordism or post-Fordism.

What they had learned nonetheless was the importance of operational learning and continuous improvement, which, ironically, had been initially "learned up" from a former manager of Ford. What they also independently had achieved was both "lean production" and "lateral learning" [WOM 90, WOM 96, COL 00].

– Thus, whether responsible for cost control, the pressing plant, quality control, the paint shop or any other part of pressing and assembly, operational managers in the plant made a practice of taking time to share understanding the implicit logic of each others' jobs.

– This not only was hand-in-hand, but mind-on-mind, getting to learn the multiple constraints, which their counterparts could face daily and the complex trade-offs that could be needed to overcome them.

– It meant more than getting "some idea" of what problems the others faced. They spent not a day but a week with other operational managers for up to 3 weeks per year.

– It involved seeing, sensing and sharing experience rather only the explicit logic of semantic learning. It broke through the implicit constraints of a Taylorist "my job–your job" demarcation by "coming to understand" in depth.

– The outcome was competitive advantage *within* the VW Group. Autoeuropa became more productive on VW's own criteria for operational efficiency than any of its German plants including the famed Wolfsburg.

– The lateral learning was to mutual advantage. But this was reinforced by a commitment to no involuntary redundancies. When a product reached the end of its market cycle, workers were given shorter working time, retraining or a combination of both.

– Despite the financial crisis from 2008, Autoeuropa gained three additional models and became a four-car plant.

Source: Site visits, one-to-one interviews, group discourse and workshops with managers and employees of Autoeuropa.

7.12. Conclusion

Operational logic may be innovative and creative, as in the case of Japanese post-Fordism, or inertial as in Sloan's multidivisional management model, which could not achieve operational learning since it presumed that it had achieved the height of innovative management through it multidivisional model. Big business may have been founded as an outcome of Schumpeter's "creative aptitude" by an innovative entrepreneur, yet then succumb to inertial organizational logic when the need to organize and systemize becomes bureaucratized. Countering also the widespread public perception that government bureaucracies necessarily are inefficient and lack innovative capacity, the evidence from Mazzucato [MAZ 11], Block and Keller [BLO 11] and Lazonick [LAZ 10] has demonstrated that government can foster Schumpeter's creative aptitude and do so in a manner that spans boundaries between both intellectual disciplines and its own departments.

The chapter has shown that while management theorists such as Porter [POR 98] have dismissed *kaizen*, and that while references to it in many management texts tend to assume that it means continuous improvement, it was the commitment to lifetime employment for core workers in Japan that enabled *kaizen* in the dual sense of the kai of improvement and the zen of mutual advantage. We have drawn implications from this, and deconstruction of meta-analyses of lifetime employment in Japan for theories of psychological and social contract and their relations to both explicit and implicit institutional logics.

We also have indicated that the presumption during the Eurozone crisis of the southern periphery of the EU as uncompetitive is not supported in the case of Volkswagen's Autoeuropa subsidiary in Portugal, which has outstripped all of VW's German plants in terms of Volkswagen's own internal performance criteria. We suggest that there are implications from this for European innovation technology policy. For example, the Science and

Technology Framework Programmes of the European Commission are not proactive in the sense of those pursued in the United States but depend on submissions independently suggested by research departments of companies, or universities or research institutes. Yet, they also lack institutional follow-through. Most of them that succeed are signed off without creating a new high-tech start-up or even a new research contract. They also are gripped by a misplaced institutional logic that the criteria for approval for such projects are not that they should create promote entrepreneurship but that the submission should be considered publishable in a peer-reviewed journal, which is self-referential academic recognition rather than promoting and sustaining process or product innovation.

7.13. Future research directions

The chapter implies that future research into institutional logics and their relevance for enabling or inhibiting innovation could more clearly distinguish between organizational and operational logics and between organizational and operational learning. One of the main directions for such research could be to assess the degree to which most process innovation and learning is at operational rather than organizational levels, and not only that upper management needs to "delve down to learn up" from this, but also that lateral learning in a non-hierarchical manner at operational levels can enable innovation, as found by Mørk *et al.* [MØR 12], in a public institution such as a hospital, and indicated by the findings of Block, Lazonick [LAZ 10] and Mazzucato [MAZ 11] in symbiosis between public and private sector institutions. A related area for research could be a deeper assessment of the degree to which European innovation has been constrained both by the degree to which its science and technology policies have been passive rather than proactive, and the lessons that can be learned for this from the success of synergic boundary spanning.

Future sociocognitive research on institutional logic might also gain from more central recognition of less than conscious Kleinian projective identification, displacement and denial, such as that which may inhibit further "small step" innovation even after a "big leap" change such as that achieved by Ford with the integrated assembly line for the Model T. Future research also might relate institutional logics to the "insider–outsider" debate and the case for more flexible labor markets, which has been known to claim that Europe, including Germany, needs to reduce "insider" employee rights and allow "outsider" competition for jobs [LIN 88, SIN 03]. For as Moriguchi and Ono [MOR 04] have shown, and although not elaborated upon in this chapter, it was precisely competition for trained engineers in Japan that led to most of its engineering firms adopting lifetime employment and profit sharing to increase "insider" attachment.

7.14. Bibliography

[ARG 74] ARGYRIS C., SCHÖN D.A., *Theory in Practice*, Jossey-Bass, San Francisco, CA, 1974.

[ARG 78] ARGYRIS C., SCHÖN D.A., *Organisational Learning: A Theory of Action Perspective*, Addison-Wesley, Reading, MA, 1978.

[ARG 96] ARGYRIS C., SCHÖN D.A., *Organisational Learning II*, Addison-Wesley, Reading, MA, 1996.

[BLA 04] BLACK L., CARLILE P., REPENNING N.P., "A dynamic theory of expertise and occupational boundaries in new technology implementation: building on Barley's study of CT scanning", *Administrative Science Quarterly*, vol. 49, no. 4, pp. 572–607, 2004.

[BAR 06] BARLEY S.R., "What we know (and mostly don't know) about technical work", in ACKROYD S., BATT R., THOMPSON P., TOLBERT P. (eds), *The Oxford Handbook of Work and Organization*, Chapter 16, Oxford University Press, 2006, pp. 376–403.

[BLO 08] BLOCK F., "Swimming against the current: the rise of a hidden developmental state in the US", *Politics and Society*, vol. 36, no. 2, pp. 169–206, 2008.

[BLO 11] BLOCK F., KELLER M. (eds), *State of Innovation: The US government's Role in Technology Development*, Paradigm, Columbia, NY, 2011.

[BOU 77] BOURDIEU P., *Outline of a Theory of Action*, Cambridge University Press, Cambridge, 1977.

[BOU 84] BOURDIEU P., *Homo Academicus*, Editions de Minuit, Paris, 1984.

[BOU 90] BOURDIEU P., *The Logic of Practice*, Polity Press, Cambridge, 1990.

[BUC 02] BUCCI F., *Albert Kahn: Architect of Ford*, Princeton University Press, Princeton, NJ, 2002.

[CAR 10] CARLSON C., "In 2004, Mark Zuckerberg broke into a facebook user's private email account", *Business Insider*, 10 March 2010.

[COB 98] COBB R., *The French and Their Revolution*, GILMOUR D. (ed.), John Murray, London, 1998.

[COL 00] COLENSO M., *Kaizen Strategies for Organisational Change: Evolution and Revolution in the Organization*, Financial Times-Prentice Hall, London, 2000.

[CRO 04] CROSS R.L., PARKER A., *The Hidden Power of Social Networks: Understanding How Work Really Gets Done in Organizations*, Harvard Business School Press, Boston, MA, 2004.

[CSE 11] CSERE C., "It's all your fault: the DOT renders its verdict on Toyota's unintended-acceleration scare", *Car and Driver*, June 2011.

[DEE 05] DEEG R., "Path dependency, institutional complementarity, and change in national business systems", in MORGAN G., WHITELEY R., MOEN E. (eds), *Changing Capitalisms?*, Oxford University Press, Oxford, pp. 21–52, 2005.

[DEJ 47] DE JOUVENAL B., *Essai sur la politique de Rousseau. Preface to Rousseau, J-J. Du Contrat Social*, Les Éditions du Cheval Ailé, Geneva, 1947.

[DUR 52] DURKHEIM E., *Suicide: A Study in Sociology*, Routledge & Kegan Paul, London, 1952.

[EUR 00] EUROPEAN UNION, *European Council, Presidency Conclusions*, Lisbon, 2000.

[FRI 91] FRIEDLAND R., ALFORD R., "Bringing society back in: symbols, practices, and institutional contradictions", in POWELL W.W., DIMAGGIO P.J. (eds), *The New Institutionalism in Organizational Analysis*, University of Chicago Press, Chicago, IL, pp. 232–263, 1991.

[GAR 12] GARSIDE J., "Facebook founder Mark Zuckerberg sees stock tumble amid IPO lawsuits", *The Guardian*, 23 May 2012.

[GLA 12] GLANCE D., "Killing the Kodak moment… is the iPhone really to blame?", *The Conversation*, 9 January 2012.

[GUE 95] GUEST D.E., "Human resource management and performance", *Paper Presented at the Seminar, Human Resource Management and Performance*, the Netherlands, September 1995.

[GUE 98a] GUEST D.E., "Beyond HRM: commitment and the contract culture", in SPARROW P., MARCHINGTON M. (eds), *Human Resource Management: The New Agenda*, Financial Times-Pitman, London, pp. 37–51, 1998.

[GUE 98b] GUEST D.E., "On meaning, metaphor and the psychological contract: a response to Rousseau", *Journal of Organizational Behaviour*, vol. 19, pp. 673–677, 1998.

[GUE 99] GUEST D.E., "Human resource management: the workers' verdict", *Human Resource Management Journal*, vol. 9, no. 3, pp. 5–25, 1999.

[GUE 03] GUEST D.E., "Human resource management and organizational performance: are employees the missing link?", *Management and People' Seminar: Faculty of Economics*, University of Coimbra, Portugal, 2003.

[GUE 04a] GUEST D.E., "Flexible employment contracts, the psychological contract and employee outcomes: an analysis and review of the evidence", *International Journal of Management Reviews*, vols. 5/6, no. 1, pp. 1–19, 2004.

[GUE 04b] GUEST D.E., "The psychology of the employment relationship", *Applied Psychology*, vol. 53, no. 4, pp. 541–555, 2004.

[GUM 00] GUMPORT P.J., "Academic restructuring: organizational change and institutional imperatives", *The International Journal of Higher Education and Educational Planning*, vol. 39, pp. 67–91, 2000.

[GUS 01] GUSTON D.H., "Boundary organizations in environmental policy and science: an introduction", *Science, Technology and Human Values*, vol. 26, no. 4, pp. 399–408, 2001.

[HAV 97] HAVEMAN H.A., HAYAGREEVA R., "Structuring a theory of moral sentiments: institutional and organizational coevolution in the early thrift industry", *American Journal of Sociology*, vol. 102, no. 6, pp. 1606–1651, 1997.

[HOL 00] HOLLAND S., Innovation by agreement, Paper for the Portuguese Presidency of the European Council, Office of the Prime Minister, Lisbon, 2000.

[HOL 06] HOLLINGER P., WIESMANN G., "House of conflict: how rivalries have disrupted the steep climb of Airbus", *The Financial Times*, 14 November 2006.

[ICH 07] ICHIJO K., NONAKA I., *Knowledge Creation and Management*, Oxford University Press, Oxford, 2007.

[IZZ 11] IZZO P., "Top 10 dying industries", *Wall Street Journal Real Time Economics Home Page*, 28 March 2011.

[KAP 94] KAPLINSKY R., POSTHUMA A., *Easternisation: The Spread of Japanese Manufacturing Techniques to Developing Countries*, Frank Cass, London, 1994.

[KEY 36] KEYNES J.M., *The General Theory of Money, Interest and Employment*, Macmillan, London, 1936.

[KLE 32] KLEIN M., *The Psycho-analysis of Children*. Hogarth Press, London, 1932.

[KLE 52] KLEIN M., *Developments in Psychoanalysis*, Hogarth Press, London, 1952.

[KUH 62] KUHN T., *The Structure of Scientific Revolutions*, University of Chicago Press, Chicago, IL, 1962.

[LAC 87] LACEY R., *Ford: The Men and the Machine*, Heinemann and Pan Books, London, 1987.

[LAZ 10] LAZONICK W., *Sustainable Prosperity in the New Economy? Business Organization and High-Tech Employment in the United States*, Upjohn Institute, New York, 2010.

[LAZ 11] LAZONICK W., TULUM O., "US biopharmaceutical finance and the sustainability of the biotech business mode", *Research Policy*, vol. 40, pp. 1170–1187, 2011.

[LIN 88] LINDBECK A., SNOWER D.J., *The Insider-Outsider Theory of Employment and Unemployment*, MIT, Cambridge, MA, 1988.

[LOK 10] LOK J., "Institutional logics as identity projects", *Academy of Management Journal*, vol. 53, no. 6, pp. 1305–1335, 2010.

[LOU 02] LOUNSBURY M., "Institutional transformation and status mobility: the professionalization of the field of finance", *Academy of Management Journal*, vol. 45, pp. 255–266, 2002.

[LOU 07] LOUNSBURY M., "A tale of two cities: competing logics and practice variation in the professionalization of mutual funds", *Academy of Management Journal*, vol. 50, pp. 289–307, 2007.

[MAN 06] MANDELBROT B., TALEB N., "A focus on the exceptions that prove the rule", *The Financial Times*, 24 March 2006.

[MAZ 11] MAZZUCATO M., *The Entrepreneurial State*, Demos, London, 2011.

[MIN 94] MINTZBERG H., *The Rise and Fall of Strategic Planning*, The Free Press, New York, 1994.

[MIN 95] MINTZBERG H., QUINN J.B., GHOSHAL S., *The Strategy Process*, Prentice Hall, Hemel Hempstead, 1995.

[MIN 04] MINTZBERG H., *Managers, Not MBAs*, Financial Times-Prentice Hall, London, 2004.

[MIN 06] MINTZBERG H., "Community-ship is the answer", *The Financial Times*, Special Report on Business Education, 23 October 2006.

[MIT 07] MITCHELL T., "Lenova shares hit by IBM stake sale", *The Financial Times*, 7 February 2007.

[MOO 02] MOORMAN C., "Consumer health under the scope", *Journal of Consumer Research*, vol. 29, pp. 152–158, 2002.

[MOR 04] MORIGUCHI C., ONO H., Japanese lifetime employment: a century's perspective, Working Paper 205, European Institute of Japanese Studies, Stockholm, May 2004.

[MØR 12] MØRK B.E, HOHOLM T., MAANINEN-OLSSON E., *et al.*, "Changing practice through boundary organizing: a case from medical R&D", *Human Relations*, vol. 65, pp. 263–287, 2012.

[NAK 06] NAKAMOTO M., REED J., "Toyota to pass GM as biggest carmaker", *The Financial Times*, 24 December 2006.

[NHT 11] NHTSA, U.S. Department of Transportation Releases Results from NHTSA-NASA Study of Unintended Acceleration in Toyota Vehicles, 8 February, Washington, DC, 2011.

[NIC 11] NICOLINI D., "Practice as the site of knowing. Insights from the field of telemedicine", *Organization Science*, vol. 22, no. 3, pp. 602–620, 2011.

[NON 94] NONAKA I., "A dynamic theory of organisational knowledge creation", *Organisation Science*, vol. 5, no. 1, pp. 14–37, 1994.

[OLI 00] OLIVEIRA T.C., "Implicit logic in unstructured interviewing", *Selection and Development Review*, British Psychological Society, London, vol. 16, no. 2, pp. 10–14, 2000.

[OLI 01] OLIVEIRA T.C., "Power in panel interviewing", *Selection and Development Review*, British Psychological Society, London, vol. 17, no. 3, pp. 7–10, 2001.

[OLI 02] OLIVEIRA T.C., "Procedural justice in selection", *Selection and Development Review*, British Psychological Society, London, vol. 18, no. 2, pp. 3–7, 2002.

[OLI 07a] OLIVEIRA T.C., "Delving down to learn up", *Plenary Presentation to the Sloan – MIT Portugal Conference on New Developments in Management*, Lisbon, June 2007.

[OLI 07b] OLIVEIRA T.C., HOLLAND S., *Innovation-by-Agreement: Activating the Lisbon Agenda. Notas Economicas*, Faculty of Economics, University of Coimbra, Portugal, December 2007.

[PAN 12] PAN J., "Say goodbye to record stores and physical albums", *Infographic*, 24 July 2012.

[PAR 09] PARETO V., *Trattato Di Sociologia Generale*, Boringhieri, Rome, 1909. [Translated as *The Mind and Society*, Harcourt, Brace, New York, 1935.]

[PAR 51] PARSONS T., *The Social System*, The Free Press, New York, 1951.

[PAR 56] PARSONS T., "Suggestions for a sociological approach to the theory of organizations", *Administrative Science Quarterly*, vol. 1, pp. 63–85, 1956.

[PIL 04] PILLING D., "Canon sticks to jobs-for-life", *The Financial Times*, 10 November 2004.

[PIO 84] PIORE M.J., SABEL C.F., *The Second Industrial Divide*, Basic Books, New York, 1984.

[POL 58] POLANYI M., *Personal Knowledge*, University of Chicago Press, Chicago, IL, 1958.

[POL 62] POLANYI M., "Tacit knowing: its bearing in some problems of philosophy", *Review of Modern Physics*, vol. 34, no. 4, pp. 601–616, 1962.

[POL 68] POLANYI M., *The Tacit Dimension*, Routledge, London, 1968.

[POP 45] POPPER K., *The Open Society and Its Enemies*, 2 vols, Routledge, London, 1945.

[POR 80] PORTER M.E., *Competitive Strategy*, Free Press, New York, 1980.

[POR 90] PORTER M.E., *The Competitive Advantage of Nations*, Free Press, New York, 1990.

[POR 98] PORTER M.E., *On Competition*, Harvard Business School Press, Boston, MA, 1998.

[REA 09] REAY T., HININGS C.R., "Managing the rivalry of competing institutional logics", *Organization Studies*, vol. 30, no. 6, pp. 629–652, 2009.

[RIC 89] RICHARDS B., *Images of Freud: Cultural Responses to Psychoanalysis*, Dent, London, 1989.

[ROU 62] ROUSSEAU J.-J., *Du Contrat Social*, Garnier, Paris, 1762. (Republished 1961)

[ROU 89] ROUSSEAU D.M., "Psychological and implied contracts in organizations", *Employee Responsibilties and Rights Journal*, vol. 2, pp. 121–139, 1989.

[ROU 93] ROUSSEAU D.M., PARKS J.M., "The contracts of individuals and organizations", in CUMMINGS L.L., STAW B.W. (eds), *Research in Organizational Behavior*, vol. 15, JAI Press, Greenwich, CT, pp. 150–166, 1993.

[ROU 95] ROUSSEAU D.M., *Psychological Contracts in Organizations: Understanding Written and Unwritten Agreements*, Sage, Newbury Park, CA, 1995.

[ROU 97] ROUSSEAU D.M., TINSLEY K., "Human resources are local: society and social contracts in a global economy", in ANDERSON N., HERRIOT P. (eds), *Handbook of Selection and Appraisal*, Wiley, London, pp. 167–214, 1997.

[ROU 98] ROUSSEAU D.M., "The problem of the psychological contract", *Journal of Organizational Behavior*, vol. 19, pp. 665–671, 1998.

[ROU 07] ROUBINI N., *The Black Swan: The Impact of the Highly Improbable*, Random House, New York, 2007.

[ROT 03] ROTHENBERG S., "Knowledge content and worker participation at NUMMI", *Journal of Management Studies*, vol. 40, no. 7, pp. 1783–1802, 2003.

[SCH 49] SCHUMPETER J., *The Theory of Economic Development*, Harvard University Press, Cambridge MA, 1949.

[SCH 75] SCHNEIDER M., *Neurosis and Civilization*, Seabury Press, New York, 1975.

[SCO 81] SCOTT W.R., "Boundary setting and boundary spanning", in SCOTT W.R. (ed.), *Organizations: Rational, Natural, and Open Systems*, Prentice Hall, Englewood Cliffs, NJ, pp. 179–206, 1981.

[SCO 00] SCOTT W.R, RUEF M., MENDEL P., *et al.*, *Institutional Change and Health Care Organizations: From Professional Dominance to Managed Care*, University of Chicago Press, Chicago, IL, 2000.

[SIN 03] SINN H.-W., The Laggard of Europe, *CESifo Forum*, vol. 4, no. 1, pp. 1–32, 2003.

[SMI 59] SMITH A., *The Theory of Moral Sentiments*, RAPHAEL D.D., MACFIE A.L. (eds.), Oxford University Press, Oxford, 1759.

[SPE 12] SPECTOR M., MATTIOLI D., "Kodak teeters on the brink", *The Wall Street Journal*, 5 January 2012.

[STE 06] STERN S., "The problem with performance – managing professionals", *The Financial Times*, 11 July 2006.

[SWA 02] SWAN J., SCARBROUGH H., ROBERTSON M., "The construction of 'communities of practice' in the management of innovation", *Management Learning*, vol. 33, no. 4, pp. 479–496, 2002.

[SWA 05] SWAN J., SCARBROUGH H., "The politics of networked innovation", *Human Relations*, vol. 58, no. 7, pp. 913–943, 2005.

[SWA 07] SWAN J., BRESNEN M., NEWELL S., *et al.*, "The object of knowledge: the role of objects in biomedical innovation", *Human Relations*, vol. 60, no. 12, pp. 1809–1837, 2007.

[TAL 61] TALMON J.L., *The Origins of Totalitarian Democracy*, Heinemann, London, 1961.

[TAY 11] TAYLOR F.W., *The Principles of Scientific Management*, Harper and Row, New York, 1911.

[THO 99] THORNTON P.H., OCASIO W., "Institutional logics and the historical contingency of power in organizations: executive succession in the higher education publishing industry, 1958–1990", *American Journal of Sociology*, vol. 105, no. 3, pp. 801–843, 1999.

[THO 02] THORNTON P.H., "The rise of the corporation in a craft industry: conflict and conformity in institutional logics", The Academy of Management Journal, vol. 45, no. 1, pp. 81–101, 2002.

[THO 05] THORNTON P.H., JONES C., KURY K., "Institutional logics and institutional change: transformation in accounting, architecture, and publishing", in JONES C., THORNTON P.H. (eds), *Research in the Sociology of Organizations*, JAI, London, pp. 125–170, 2005.

[THO 08] THORNTON P.H., OCASIO W., LOUNSBURY M. (eds), *The Institutional Logics Perspective*. Oxford University Press, 2012.

[WAT 06] WATERS R., "The wizardry of woz", *The Financial Times*, 30 September–1 October 2006.

[WEB 04] WEBER M., *The Theory of Social and Economic Organisation*, PARSONS T. (ed.) (1930), William Hodge, London, 1904.

[WOM 90] WOMACK J.P., JONES D.T., ROOS D., *The Machine that Changed the World*, Harper Collins, New York, 1990.

[WOM 96] WOMACK J., JONES D.T., *Lean Thinking*, Simon and Schuster, New York, 1996.

[WOM 05] WOMACK J.P., JONES D.T., *Lean Solutions: How Companies and Customers Can Create Wealth Together*, Simon Schuster, New York, 2005.

[YAM 04] YAMAGUCHI K., "The state and change in 'lifetime employment' in Japan: from the end of war through 1995", *Keizai Sangyo Journal, Research Institute of Economy, Trade and Industry, Research Review*, vol. 11, 2004.

[ZAJ 04] ZAJAC E.J., WESTPHAL J.D., "The social construction of market value: institutionalization and learning perspectives on stock market reactions", *American Sociological Review*, vol. 69, pp. 433–458, 2004.

[ZHO 05] ZHOU X., "The institutional logic of occupational prestige ranking", *American Journal of Sociology*, vol. 111, no. 1, pp. 90–140, 2005.

7.15. Further recommended reading

[ABO 07] ABO T. (ed.), *Japanese Hybrid Factories: A Comparison of Global Production Strategies*, Macmillan Palgrave, Houndsworth, 2007.

[BEA 98] BEARDWELL I., "Bridging the gap? Employee voice, representation and HRM", in SPARROW P., MARCHINGTON M. (eds), *Human Resource Management: The New Agenda*, Financial Times–Pitman Publishing, London, 1998.

[BEC 94] BECK J.C., BECK M.N., *The Change of a Lifetime Employment Patterns among Japan's Managerial Elite*, University of Hawaii Press, Honolulu, 1994.

[BER 89] BERGGREN C., "New production concepts in final assembly", in WOOD S. (ed.), *The Transformation of Work*, Unwin Hyman, London, pp. 172–203, 1989.

[BLA 04] BLACK L., CARLILE P., REPENNING N.P., "A dynamic theory of expertise and occupational boundaries in new technology implementation: building on Barley's study of CT scanning", *Administrative Science Quarterly*, vol. 49, no. 4, pp. 572–607, 2004.

[COL 71] COLE R.E., *Japanese Blue Collar – The Changing Tradition*, University of California Press, Berkeley, CA, 1971.

[COL 79] COLE R.E., *Work, Mobility and Participation: A Comparative Study of American and Japanese Industry*, University of California Press, Berkeley, CA, 1979.

[CON 05] CONWAY N., BRINER R., *Understanding Psychological Contracts at Work*, OUP, Oxford, 2005.

[DOR 73] DORE R., *British Factory–Japanese Factory*, Alden Press, Oxford, 1973.

[DOR 86] DORE R., *Flexible Rigidities and Industrial Policy in the Japanese Economy*, The Athlone Press, London, 1986.

[DRU 71] DRUCKER P.F., "What we can learn from Japanese management", *Harvard Business Review*, March-April, pp. 110–122, 1971.

[EZZ 01] EZZAMEL M., WILMOTT H., WORTHINGTON F., "Power, control and resistance in 'The factory that time forgot'", *Journal of Management Studies*, vol. 38, no. 8, pp. 1053–1079, 2001.

[FOR 23] FORD H., *My Life and Work*, Doubleday, Garden City, NY, 1923.

[GEN 00] GENDA Y., REBICK M.E., "Japanese labour in the 1990s: stability and stagnation", *Oxford Review of Economic Policy*, vol. 16, pp. 85–102, 2000.

[GRA 75] GRAMSCI A., *Americanismo e fordismo, in Note sul Machiavelli*, Editori Riuniti, Rome, pp. 401–444, 1975.

[HEL 03] HELLER F., "Worker participation in change: a critical assessment", *Applied Psychology*, vol. 41, 6, pp. 144–163, 2003.

[ICH 07] ICHIJO K., NONAKA I., *Knowledge Creation and Management*, Oxford University Press, Oxford, 2007.

[ING 10] INGRASSIA P., *Crash Course: The American Automobile Industry's Road from Glory to Disaster*, Random House, New York, 2010.

[JAP 09] JAPAN PRODUCTIVITY CENTER FOR SOCIAL-ECONOMIC DEVELOPMENT, White Paper on Mental Health of Workers, Tokyo, 24 August 2009.

[KON 84] KONO T., *Strategy and Structure of Japanese Enterprises*, Macmillan, London, 1984.

[KOS 83] KOSHIRO K., Personnel planning, technological changes, and outsourcing in the Japanese automobile industry, Studies, Discussion Paper Series 83–3, Yokohama National University, Center for International Trade, May 1983.

[LAZ 10] LAZONICK W., *Sustainable Prosperity in the New Economy? Business Organization and High-Tech Employment in the United States*, Upjohn Institute, New York, 2010.

[LAZ 11] LAZONICK W., TULUM O., "US biopharmaceutical finance and the sustainability of the biotech business mode", *Research Policy*, vol. 40, pp. 1170–1187, 2011.

[LOC 94] LOCKE J., *Treatise Concerning the True Origin, Extent and End of Civil Government*, 1694. [Republished in Barker E. (ed.), *Social Contract*, OUP, Oxford, 1960.]

[MON 03] MONIN N., BARRY D., MONIN D.J., "Toggling with Taylor: a different approach to reading a management text", *Journal of Management Studies*, vol. 40, no. 2, pp. 377–402, 2003.

[MOT 11] MOTOYAMA Y, APPELBAUM R., PARKER R., "The national nanotechnology initiative: federal support for science and technology, or hidden industrial policy?", *Technology in Society*, vol. 33, pp. 109–118, 2011.

[NAK 07] NAKAMOTO M., REED J., "Toyota's record-profits run to continue", *The Financial Times*, 7 February 2007.

[NON 95] NONAKA I., TAKEUCHI H., *The Knowledge-Creating Company*, Oxford University Press, New York, 1995.

[OHM 82] OHMAE K., *The Mind of the Strategist: The Art of Japanese Business*, McGraw-Hill, New York, 1982.

[OHN 88a] OHNO T., *The Toyota Production System*, Productivity Press, Portland, OR, 1988.

[OHN 88b] OHNO T., *Workplace Management*, Productivity Press, Portland, OR, 1988.

[PIL 04] PILLING D., "Canon sticks to jobs-for-life", *The Financial Times*, 10 November 2004.

[PIL 06] PILLING D., "Cash in the kaisha: how an overhaul is producing a resurgent corporate Japan", *The Financial Times*, 23 March 2006.

[QUI 95] QUINN J., "The Honda motor company", in MINTZBERG H., QUINN J.B., GHOSHAL S. (eds), *The Strategy Process*, Prentice Hall, London, pp. 297–318, 1995.

[QUI 09] QUINN J., "GM files for Chapter 11 bankruptcy protection", *The Telegraph*, 1 June 2009.

[REB 01] REBICK M.E., "Japanese labour markets: can we expect significant change?", in BLOMSTROM M., GANGNES B., LA CROIX S. (eds), *Japan's New Economy: Continuity and Change in the Twenty-First Century*, Oxford University Press, Oxford, pp. 120–141, 2001.

[ROU 62] ROUSSEAU J.-J., *Du Contrat Social*, Garnier, Paris, 1762. (Republished [1961])

[SAK 97a] SAKO M., "Forces for homogeneity and diversity in the Japanese industrial relations system", in SAKO M., SAKO H. (eds), *Japanese Management and Labour in Transition*, Routledge, London, 1997.

[SAK 97b] SAKO M., HIROKI S., *Japanese Labour and Management in Transition: Diversity, Flexibility and Participation*, Routledge, London and New York, 1997.

[SCH 83] SCHNEIDER B., "Interactional psychology and organisation behaviour", in STAW B.M., CUMMINGS L.L. (eds), *Research in Organisation Behaviour*, vol. 5, JAI Press, Greenwich City, pp. 1–31, 1983.

[SCH 87] SCHNEIDER B., "The people make the place", *Personnel Psychology*, vol. 40, pp. 437–453, 1987.

[SCH 94] SCHEIN E.H., "On dialogue, culture and organisational learning", *Organisational Dynamics*, vol. 22, pp. 40–51, 1994.

[SEN 90] SENGE P., *The Fifth Discipline*, Doubleday, New York, 1990.

[SHI 88] SHINOHARA I., *New Production Systems: JIT Crossing Industry Barriers*, Productivity Press, Cambridge, MA, 1988.

[SHI 92] SHIMADA H., "Japan's industrial culture and labour-management relations", in KUMON S., ROSOVSKY H. (eds), *The Political Economy of Japan, Vol. 3: Cultural and Social Dynamics*, Stanford University Press, Stanford, CA, 1992.

[SHI 06] SHIROUZU N., "Toyota turns to U.S. for road map on ramping up in Chinese market", *The Wall Street Journal*, 26–28 May 2006.

[SIM 05] SIMON B., "Ford cuts output after seeing 38% plunge in earnings", *The Financial Times*, 21 April 2005.

[SIM 06] SIMON B., "End of the road for thousands of car workers", *The Financial Times*, 23 June 2006.

[SLO 64] SLOAN A., *My Years with General Motors*, Doubleday, New York, 1964.

[SOB 09] SOBLE J., "Toyota buoyed by surge in orders for new Prius", *The Financial Times*, 19 May 2009.

[SOR 56] SORENSEN, C.E., *My Forty Years with Ford*, Norton, New York, 1956.

[STJ 97] ST. JOHN M.F., SHANKS D.R., "Implicit learning from an information processing standpoint", in BERRY D.C. (ed.), *How Implicit is Implicit Learning?*, Oxford University Press, Oxford, 1997.

[STY 01] STYHRE A., "*Kaizen*, ethics and the care of operations: management and empowerment", *Journal of Management Studies*, vol. 38, no. 8, pp. 795–810, 2001.

[TSO 01] TSOUKAS H., VLADIMIROU E., "What is organisational knowledge?", *Journal of Management Studies*, vol. 38, no. 7, pp. 973–994, 2001.

[TSU 98] TSUTSUI W., *Manufacturing Ideology: Scientific Management in Twentieth-Century Japan*, Princeton University Press, Princeton, NJ, 1998.

[UND 96] UNDERWOOD G. (ed.), *Implicit Cognition*, Oxford University Press, Oxford, 1996.

[VAL 11] VALDES-DAPENA P., "Pedals, drivers blamed for out of control Toyotas", *CNN Money*, 8 February 2011.

[WAL 85] WALTON R.E., "From control to commitment in the workplace", *Harvard Business Review*, vol. 63, no. 2, pp. 77–84, 1985.

[WAT 06] WATERS R., "The wizardry of woz", *The Financial Times*, 30 September–1 October 2006.

[WEI 96] WEICK K.E., WESTLEY F., "Organizational learning: affirming an oxymoron", in CLEGG S.R., HARDY C., NORD W.R. (eds), *Handbook of Organization Studies*, Sage, London, pp. 440–458, 1996.

[WOO 03] WOOLFE MORRISON E., MILLIKEN F., "Speaking up, remaining silent: the dynamics of voice and silence in organisations", *Journal of Management Studies Special Issue*, vol. 40, no. 6, pp. 1353–1568, 2003.

[ZAL 98] ZALEZNIK A., *Managers and Leaders: Are They Different? Harvard Business Review on Leadership*, Harvard Business School Press, Boston, MA, pp. 61–88, 1998.

Chapter 8

HRM in SMEs in Portugal: An Innovative Proposal of Characterization

This chapter aims to understand the role of human resources (HR) management in small and medium enterprises (SME) in Portugal and to contribute to a better knowledge of this area, which is still only briefly studied. More particularly, it aims to understand the level of implementation of HR practices in this type of organization.

8.1. Introduction

The economic and social crisis that we have crossed has triggered several public debates about the importance of SMEs for Portuguese business. For decades, SMEs were seen as companies without physical labor, essentially family and rudimentary fundamentals of management. Currently, the image is tending to change, mainly due to several economic crises that have plagued the country and the world, which are extinguishing less prepared companies, allowing the development of technology companies and those with other capacities to survive in the global market. The demand for competitive factors has led entrepreneurs to invest in training for their employees and invest in quality processes, encouraging formalization of all critical business processes.

Chapter written by Pedro Ribeiro NOVO DE MELO and Carolina Feliciana MACHADO.

All the important components of a company need to find ways to increase their competitiveness and efficiency. HR cannot disregard these changes. As a vital part of an organization, HR need mechanisms, processes and practices that make them more productive, motivated and committed. Then, policies and practices of human resource management (HRM) arise in order to convert HR into strategic resources or in-line with the business strategy.

The approach of HRM's impact on the performance of an organization, so far reported in literature, has relied on investigations of large companies. There are few studies that address the idiosyncrasies of small businesses. Studies related to HRM in SMEs focus on HRM practices used in large companies, but adapted to the size, resources and culture of small businesses. Only recently, authors such as Cassel *et al.* [CAS 02], Kok and Uhlaner [KOK 01] and Cardon and Stevens [CAR 04] have addressed issues related to HRM in SMEs, looking beyond larger companies.

In Portugal, the context in which we stand, the role of HRM has gone through an evolutionary process, as in other countries, which has been described and analyzed by some scholars. However, little or nothing is known about the HR function, as mentioned Cabral-Cardoso [CAB 06]. Studies on HRM are scarce and those that exist are associated with large companies. The choice of theme and its relevance are linked to some extent to low investment in research into SME management. Another interesting aspect in SMEs is that, while organizations are smaller in size, HRM may appear either informal or unsophisticated, whereas in complex and highly structured organizations, it allows the researcher to contact with a wide variety of situations.

8.2. SMEs

SMEs are relevant to almost all economies of the world, particularly focusing on economies of developing countries, and constitute a major source of entrepreneurial skills, innovation and employment. In the enlarged European Union (EU) of 25 countries, some 23 million SMEs ensure approximately 75 million jobs and represent 99% of total enterprises [IAP 05].

Like any other sector of an economy, the size and importance of SMEs varies from country to country. In developing countries with an economy of small scale and micro-business sectors, SMEs constitute a large share of the business, which explains their strategic importance. Developing countries without a strong industry in terms of SMEs tend not only to have financial capital concentrated in large companies, but also to have elite workers that absorb much of capital of that country's economy. SMEs assume an important role in balancing the social and cultural life of a nation. However, it is not only developing countries that assume this role. According to Audretsch and Thurik [AUD 01], globalization reduced the ability of economies of Western Europe and North America in growth and job creation. It makes sense that there is a focus on developing a strong SME sector, which largely supports innovation and employment in a country. The tendency is that in the future, job creation and new types of employment will be dominated by SMEs, which can improve the economy and act as the main source of innovation. Innovation policies aimed for these companies can be an instrument for stimulating growth and competitiveness of nations. SMEs tend to dominate new and fast-growing industries. Economies that discourage SMEs are therefore likely to discourage industries with new dynamics and new products and/or services.

The SME sector includes companies in all major types of economic activity. In Portugal, SMEs are associated with 16 main sectors of activity [IAP 05]:

– fishing and aquaculture;

– extractive industries;

– manufacturing;

– electricity;

– water;

– construction;

– trade;

– transportation and warehousing;

– accommodation and restoration;

– information and communication activities;

– real estate activities;

– consulting activities;

– administrative activities;

– education;

– human health activities;

– artistic activities;

– other service activities.

Each sectors of activity have different characteristics that cause a range of perspectives and operating logics. Despite differences related to the nature of the product/service developed, culture or structure instituted, SMEs share some characteristics, such as a simple organizational structure, proximity to market and customer and having a human capital of less than 250 employees.

The role of SMEs in the economy of nations is quite big, which requires an awareness from entrepreneurs, governments and managers for the constant improvement of management models applied to SMEs. One of the less well studied areas of management in SMEs is HRM. There has been a substantial increase in the number of studies on SMEs, but even these are based mainly on finance, marketing and operational management features.

The development of new management models applied to SMEs is crucial. Most management models are adapted from large companies, which originate from adapting the difficulties of models to the structure and culture of SMEs. A large number of SMEs suffer a significant downturn in certain moments of their organizational life, with negative implications for the safety of their employees. In part, this reflects the process of "survival of fittest" in which companies without strong entrepreneurial skills or sufficient niches do not survive. This entrepreneurship skill is, nowadays, a crucial point for the essence of SMEs, not only the entrepreneurial capacity to develop new products/services, but also in day-to-day management.

8.2.1. *Characteristics of SMEs*

There is a growing recognition of the role that SMEs play in global economic recovery. For the United Nations Environment Programme, SMEs have at least three contributions to the world economy [UNE 06]. The first refers to the creation of new jobs. The second contribution is that SMEs are a source of considerable innovation activities, contributing to the development of entrepreneurship and competitiveness. Finally, SMEs add greater flexibility to industrial structure and promote great dynamism in economy.

According to Organization for Economic Co-operation and Development (OECD) data [OEC 04], in the EU, SMEs are socially and economically important. SMEs represent 99% of companies in EU and represent two-thirds of all workers. In Latin America, about 80–90% of companies are SMEs in regional economies, only Argentina has experienced a decrease in the number of SMEs (between 1998 and 2002), while in Brazil and Mexico there was a considerable increase. In Japan, 81% of employment is in SMEs, where average workers per firm are nine, four more than in the EU.

SMEs have a huge variety of organizational structures and cultures, making it difficult to define corporate policies. Strategy will be used as guiding principles that help companies to be responsible in decision-making. So, with the enormous diversity of structures and cultures, it becomes complex to characterize an SME. Comparing SMEs with large companies, their characteristics can be pointed out as:

– simple organizational structure;

– limitation of financial and HR;

– limited access to technology;

– proximity to market and customer;

– relative flexibility and adaptability to external environmental change.

These characteristics refer to both the logic of their organization as well as in relation to relationships with customers, suppliers and government agencies, among others. The limitation of financial and HR is, often, an obstacle to the use of more formalized HR policies.

Informality is a variable that normally characterizes SMEs, especially in their working relationships. According to Storey [STO 94] investigations into companies indicate that small organizations are better prepared to operate in a more informal and flexible way than large firms. The organizational structure of SMEs allows informality in employment relations and management practices. This happened because the manager lies at the operational part of company, which facilitates open and informal communication.

SMEs are organizations that play an important role in the global economy. The degree of growth/evolution of an SME is very large. The difference between growth/development and consolidation of company is in how some small businesses look to consolidate, such as going out of business, while others look at it as an opportunity to grow [FRA 98]. Organizations typically follow a common evolutionary route that makes relatively predictable steps of change. According to Greiner [GRE 72], this change occurs over five phases, in which relatively stable, evolution periods, are followed by periods of seizure revolution. These stages are associated with management problems. Churchill and Lewis [CHU 83] present a more inclusive model. According to the authors, firms go through five stages, namely existence, survival, success, imitation and maturity. These models allow us to recognize and understand that growth/evolution brings increasing complexity and challenges to management skills [MCE 00]. Fraza [FRA 98] states that for many fast-growing SMEs, the main problem is to seek and retain high-quality employees.

8.3. HRM

In 1970, the word "human resources" began to be used in the vocabulary of British Academies [HEN 86]. It was only in the 1980s that the concept of HRM originated in the United States [GUE 90], assuming a concept distinct in terms of philosophy and approach of staff management in organizations [CAB 99]. The onset of HRM was due to external factors such as increased competitiveness in domestic and international markets, together with the desire to understand the reason for low productivity growth in the US industry [GUE 90].

After two decades of changes in management practices introduced by HRM, there is evidence that many companies see their HR as valuable for business and as a key to their competitive advantage [CAB 04]. HRM, assuming a leading role with regard to management and moving away from a purely reactive, operational and administrative model, emerges as an evolutionary step up and bursting with personnel management (PM).

8.3.1. *HRM versus personnel management*

Evolution in people management was not always peaceful. The route to building a management model originated in differences between two paradigms: PM and HRM. The search for mechanisms to increase productivity and quality led to changes in how they inter-relate to management and employees [GUE 90]. Demand for more efficient and effective ways to deal with employees triggered new theories about managing people. These two models differ. According to Guest [GUE 87], the main differences between PM and HRM arise in the following variables: "manpower planning", where the PM is reactive and HRM is proactive; "psychological contract" of constraint in PM and involvement in HRM; "external control" in PM and "self-control" in HRM; "collectivist relationship" in PM and "individual" in HRM and finally "cost minimization" in PM to "maximize the investment" in HRM.

The emergence of the concept of HRM had four periods: a traditional/conservative period (period of administrative efficiency and demand reduction of personnel costs), radical/conservative period (period statement of management in a context where workers were seen as disposable), pluralist/innovative period (emergence of a new industrial relation that seeks a cooperative relationship between management and unions) and unitarian/innovative period (based on values and policies that take advantage of organization talents [GUE 90]). Evolution enabled the concept to introduce a more human component oriented towards talent and performance.

This people management model does not lead by consensus. For Guest [GUE 89], this model may not be the most feasible for organizations. However, what should be noted is that despite the criticism, HRM is emerging in worldwide organizations. The truth is that for some authors (e.g. [TOR 89, ARM 00]), this trigger is the result of the natural adaptation of PM to the current business context, "labeled" with something more persuasive and attractive.

8.3.2. *HRM in Portugal*

For decades, Portugal has had a significant evolution in the search for continuous improvement of its HRM policies and practices. To understand the existing evolution an analysis in a macrostructural perspective is needed, contrasting it with more local perspective. The main theoretical models of HRM emerged in the United States. The debate that has arisen and could be considered to this theoretical analysis is the transition of these theoretical models, developed based on American organizations, to organizations around the world.

There are several authors who analyze the European space through clusters (e.g. [NIK 05, STA 05, BRE 07]). Researchers have developed models to explain the role of HRM on comparative perspectives of different countries [BRE 04]. This perspective based on cluster proves the existence of multiple overviews in what concerns the development of HRM in Europe. Portugal, according to researchers, is included in the Latin cluster. Belgium, Italy, Spain and France are also part of this cluster.

There are identifiable differences between how HRM is analyzed in Europe and its situation in the United States. The European perspective provides various notions of "American" HRM [BRE 93]. Portugal has its particularities in the management of people. Nevertheless, studies of HRM in Portugal are rare. According to Caetano and Vala [CAE 07], available data on HRM in Portugal and results from activities undertaken by APG organization (Portuguese Association of Technical Managers and Human Resources) are scarce.

The Portuguese business sector is composed, essentially, of SMEs. These companies play an important role because they generate wealth and employment [DUA 02]. For Cabral-Cardoso [CAB 06], this reality limits the development of HRM in Portugal. According to the author, the structural conditions of economy, mainly composed of SMEs and family businesses, without a personnel function, do not seem to be a favorable environment for the growth of HRM in Portugal. When we try to identify whether organizations have an HR director, size seems to be a crucial factor. At least 73% of all HR directors are embedded in medium and large firms [CAB 04].

Portugal has been experiencing a transition phase over the last three decades. The country experienced, during the 20th Century, a totalitarian regime, which had the motto "proudly alone" [CUN 05]. According to Cabral-Cardoso [CAB 06], Portugal until 1960 can be described as a poor society, essentially agrarian, ruled by a political elite from a small number of well-established families. After that, in the phase of political and social turmoil experienced in the 1960s and 1970s, Portugal faced major changes in management and managers. During these three decades, Portugal underwent profound changes in its society and business. Regarding the role and status of HR managers, in recent decades, we did not face such major changes [CAB 04].

Despite significant changes, the role of HRM in Portugal remains essentially bureaucratic and administrative [CAB 04]. Administrative guidance in managers results from a large public state intervention, a bureaucratic apparatus and a lack of customer orientation [CUN 05]. In Portugal, HRM is still at an early stage in its development and implementation. HR managers also seek a status within an organization that enable them to have a more strategic role in management decisions [CAB 04].

In recent years, with the emergence of academic backgrounds related to the theme of HRM, we have observed some changes in Portugal. According to Cabral-Cardoso [CAB 06], academic training in HRM provided a status to the academic field of HR and managers, which are better prepared for Portuguese organizations. The existence of managers with high levels of qualification in Portuguese organizations allowed a better adaptation of American rhetoric to Portuguese reality [CAB 06]. According to Cunha [CUN 05], some Portuguese managers are apparently trying to combine the best practices in local contexts.

Organizations in Portugal are beginning to identify their HR as an important and competitive advantage for their business [CAB 04]. Despite all these constraints and limitations to the development of HRM in Portugal, there have been progresses in implementing practices that enhance human resource development in organizations. Organizations are starting to include an HR department in their organizational structure, allowing more centralized management practices, greater credibility of HR managers, the manager's vision as an expert in the field of people management [CAB 06], as well as a better integration of their strategy, greater influence at the top and HR policies aligned with business [CAB 04].

8.4. HRM in SMEs

Until recently, HRM was considered not to apply to small businesses. The small amount of research and studies on SMEs [REI 01] has contributed to this. Lately, research on HRM practices in SMEs has increased considerably [KOK 01]. One possible explanation for this phenomenon is a major source of material and information from studies on SMEs that provides researchers in the field of management with the development of new theoretical currents [CAS 02]. Several authors have suggested, in their theories and empirical research, that HR can be a source of competitive advantage (e.g. [BAR 98, CUN 07]). HRM has an important role in practical application that benefits collaborators both in performance and motivation. In most cases, HRM practices benefit the progress and modernization of SMEs [MCE 84]. A study in the United States proved that poor emphasis on HR in SMEs was one of several reasons for businesses to have failed in the analyzed companies [MCE 84].

SMEs have difficulties in implementing policies and practices of HRM. For example, a study by Heneman and Berkley [HEN 99] confirmed that in 117 small businesses, only 15 had an HR department. One reason for this is the difficulty in recruiting and retaining employees due to scarce financial resources and a reluctance to lead with great cost or prohibitive practices. According to Cardon and Stevens [CAR 04], in SMEs where resources are scarce, the probability of having an HRM department with professionals is very small.

Defining HRM in SMEs is influenced by two things: informality in small firms versus formality in large companies and HRM practices of large companies tailored to small businesses. Most small businesses use informal HRM practices. Defining HRM in SMEs involves a comparison with HRM in large companies. Most studies on HRM are developed for large companies. The application of HRM practices in small businesses generally includes practices used in large companies, but adapted to size, resources and the culture of small businesses. Employment relations in SMEs can be complex, contradictory and informal, rather than harmonious or autocratic [RAM 91]. In SMEs, employees tend to develop a number of different jobs, unlike workers in large firms [KOK 01].

HRM practices in SMEs are often defined as informal. There are contextual variables that predict the existence of HRM practices in firms and their degree of formality. According to Jackson *et al.* [JAC 89], variables are firm size, strategy, technology and structure. Until 1970, the main HRM research focused on development of effective HRM practices in large organizations, studies that related HRM practices and performance of HR [KOK 01]. After that, other studies centered their attention on the impact of more formal HRM practices on organizational performance level (e.g. [HUS 95, MAC 95]). Research by these authors analyzed the impact of innovative HRM practices when integrated with organizational policies. They advocate the model of the "best practices" [KOK 01].

In general, firm size is related to incidence in planning of HRM and the degree of formalization of HRM (e.g. [KOT 04, KOK 01]). Despite the size of companies influencing HRM practices, it is not clear that it is the only important factor, given the variation in the formalization of HRM practices [KOK 01]. An empirical study by Kok and Uhlaner [KOK 01] proved the importance of firm size as an indication of a formal development of HRM. Moreover, SMEs vary the types of HRM practices. Another important aspect of the study is that the main influence of HRM practices is resource assessment rather than the expectations of stakeholders, that is, investment and planning of HR are positively associated with labor productivity.

Several studies on SMEs take them as informally organized and with informal HRM practices. However, these features are not unanimous. A perspective that has emerged shows that there are many exceptions to this model. Outside the variable size of the company there are other contextual factors that may influence HRM practices [KOK 01]. Probably, looking at the formal and informal mechanisms through which SMEs manage their workers, it is observed that these will provide the best theoretical and empirical perspective of practices of SMEs, as well as how synergistic approaches contribute to the development of HR [CAR 04].

SMEs, over the years, have been sidelined in the main studies examining HRM. These studies had, as a starting point, the principle that HRM practices developed in large companies were the same in small businesses. But this perspective is not entirely correct. SMEs have specific characteristics that do not allow a simple adjustment of policies and practices of large firms. It is true that SMEs, with links to a large company, are better able to implement certain HRM practices [KOK 01], but this is because they share resources and organizational policies. Barber *et al.* [BAR 99] suggested that using only existing models of HRM for SMEs is not significant. HRM practices of large companies have not had positive results (authors report that recruitment in large companies is more formal and bureaucratic than in SMEs). Instead, researchers should focus on specific theories and practices for SMEs and strategically focus on multiple characteristics, not only the company, but also the employees [CAR 04]. According to Cardon and Stevens [CAR 04], there is not necessarily a right way of HRM that SMEs can adopt. For authors, researchers should explore the interaction between HRM practices and other factors, and their influence on individual and organizational performance. Barney and Wright [BAR 98] suggest that the major challenge for HRM is to develop systems of practices that create synergy effects in order to develop independent sets of best practices.

SMEs are typically associated with family businesses. This relationship is not always an easy communion. Family perspective is derogatory in nature, that is, family businesses are seen as dysfunctional with a low capacity for innovation and growth. For Ram and Holliday [RAM 93], SMEs are saturated with the ideology of family. The concept of family is crucial to understand all of the social relations in SMEs, but it is much more complex and contested than usually portrayed. For Wilkinson [WIL 99], the family style is actually a form of authoritarianism with few challenges to the company manager. Flexibility is most similar to instability where there are few procedures or systems to develop the work. Family communication is seen as a resource and as a constraint to management through flexibility. Family businesses have increased problems due to family ties in existing collaborators. This aspect causes problems in business management. According to Holland and Boulton [HOL 84], the relationship between company management and family increases organizational complexity and management problems. Managers of family business should consider their position in relation to the family but also their role in the business. The specifics of family businesses, as well as limited organizational capacity, do not allow for rapid development in the implementation of policies and practices of HRM [REI 01].

A study by Duberley and Walley [DUB 95], in 16 companies, showed that most had not adopted a line of strategic HRM. Most practices are adopted from a traditionalist perspective of HRM. The traditionalist theory of HRM, discussed in some academic articles, rarely occurs in practice. Probably, it serves as a way to promote the legitimacy of management actions, trying to show scientific management when in fact organizations worldwide are more a state of complete disorder [DUB 95]. Evidence shows that HRM practices tend to be *ad hoc* regularly in small organizations [WAG 98]. A study by Cassel *et al.* [CAS 02] suggests that there is considerable diversity in how HRM practices are implemented in SMEs and their relative success. Some HRM practices are used more than others, such as recruitment and selection. Clearly, these practices have few implications in terms of resources. The study by Cassel *et al.* [CAS 02] appeared as a model of factors that influence the application of HRM practices in SMEs. To the authors, it is necessary to diagnose the specific HRM. Specifics are influenced by a number of factors, all related to the context of current business priorities [CAS 02]. Cassel *et al.* [CAS 02] concluded in their study that given the diversity of SMEs, it seems that their managers, rather than using a coherent strategy based on implementation of HRM, are using a contingency perspective, something that cannot combine or mix.

The main theories about "best practices" and high performance management systems still neglect small organizations [WIL 99]. However, they are beginning to have more empirical studies on SMEs, which, according to Baron [BAR 03], only benefits the field of HRM. SMEs provide a context of value of single business and enhancer to test principles and theories of HRM. To conclude, if there is a compelling financial support by local investors for private initiatives, then there is clearly an easier way for such practices to be more attractive to SMEs [CAS 02].

8.5. Methodology

Aiming to characterize HRM in SMEs, we used a questionnaire sent by email to a total of 512 companies. Study seeks to contribute to a better understanding of the real situation of HRM in Portuguese SMEs, that is:

– to identify HRM policies and practices of companies, that is HR planning, recruitment and selection, training and development, performance evaluation, awards and rewards;

– to investigate the relationship between HRM and business strategy;

– to assess the degree of implementation of HRM in SMEs.

8.5.1. *Sample*

A sample was selected from the perspective of convenience from a database (target population) on the market. The target population of the study was the "1000 Best SMEs" issue 284, December 2007, from *Exame magazine*. After collection and dissemination of emails, the number of companies was reduced to 512. This occurred as many companies did not have email contact or if they had, we could not access it (error message). Table 8.1 summarizes the sample construction steps.

Total sample firms	1,000
Total emails sent	730
Invalid emails – sent by the system	149
Invalid emails – deactivated	69
Total valid emails	512

Table 8.1. *Sample*

8.5.2. *Data collection*

After a literature review, the next step was to build an instrument that would allow collection of the data necessary for research, applied to a vast number of respondents. In this sense, the instrument used for data collection was the questionnaire survey.

The questionnaire survey consists of eight parts with the following contents: organization and strategy of the company, practices of HRM in the company, recruitment and selection, training and development, performance, rewards, data relating to company and respondents.

8.5.3. *Response rate*

Total number of questionnaire surveys collected from companies was 50. This corresponds to 9.8% of the 512 companies. This is a lower number for this type of research looking to find similarities between a broad universe of companies.

Questionnaire surveys sent	512	100%
Questionnaire surveys collected	50	9.8%

Table 8.2. *Comparative table of surveys sent and collected*

In the survey collection, we have observed that some companies were associated with each other, and so the survey response reflected the reality of both. In this context, the percentage previously presented suffered a change that is not reflected. Taking this observation into account and accounting these companies, the response rate increased to 10.1%.

This percentage, although low, lies within acceptable limits for such surveys in this area. It is a value of acceptable response rate for this type of study. It is true that the higher the response rate, the better for a systematic analysis of data [ROT 98]. However, Dennis [DEN 03] states that one-third of research on SMEs in scientific journals between 1991 and 1995 had a response rate below 25%.

8.6. Results

In this chapter, we intend to analyze the results obtained in the study of HRM in Portuguese SMEs, presented in previous chapter, in light of theory and empirical studies in HRM, as well as in light of objectives and questions that guided this work. The results will support the proposed characterization.

8.6.1. *Relationship between strategies of HRM and overall business strategy*

Companies need innovative processes that allow them an advantage over their competitors. This happens in the use of technology, in planning and production, and also in HR practices. None of the respondents stated that policies and practices of HRM are not inter-related with overall business strategy. All respondents reported that there is a relationship between HR strategy and business strategy.

8.6.2. *Process term planning*

In long-term planning, companies value people. A total of 44% of the companies describe people as being important in long-term planning and 36% consider that it is very important. Only capital (46%), technology (58%) and business strategy (70%) have higher values.

8.6.3. *Priorities with employees with low pay/qualifications*

SMEs are often labeled as companies that employ mostly people with low skills and therefore with low wages. The item that companies give less importance is special needs and literacy. Priorities for businesses are productivity, customer-oriented service and the need for basic training. If we analyze the importance of various aspects related to this type of physical labor, we find that a short-term view prevails. Productivity and customer-oriented service are variables that are based on a perspective of immediate profits and demand mechanisms of seduction and retention of customers. The item that characterizes SME labor force (low-skilled) has a relative importance. Education and training do not allow short-term results. As the labor force in SMEs is so volatile, it does not allow collect, by firms, the benefits of betting on training/qualification, at least in the short term.

8.6.4. *Department/employee responsible for HR*

According to Cabral-Cardoso [CAB 04], organizations are beginning to adopt the term HRM, especially when aspects of HR are associated with a company's strategy. This study demonstrates that the emergence of HR departments in companies is granted. Most companies have an HR department. Administrators have a role in managing people. At least 28% of companies presented as responsible for HR, the administrator (CEO).

8.6.5. *HR manager role in a company*

The role of HR manager is not always clearly defined in companies. In an attempt to understand the perception that the companies responsible have on HRM, when questioned about how the role of the HR manager today might be defined and how it might be defined in the future, respondents identified the HR manager, currently, as a strategic partner and administrative specialist, both with 16%. These results are ambiguous because they convey opposite sides of an evolution of HR concept. From the literature review, we found that HRM has evolved from a purely administrative function to a greater strategic integration and greater influence with top management. Respondents appear to reflect a perspective of "political correctness", which reflects current administrative practices, a strategic perspective that result from new US and European management trends. For the future, respondents give less importance to the image of an administrative specialist (4%) and identify the HR manager as a strategic partner (22%), change agent (10%) and business partner (8%).

8.6.6. *HRM in a company*

Practices with greater involvement of HR managers are more traditional practices such as recruitment and selection (84%), performance evaluation (66%), training and development with values between 64% and 80% (training diagnostics, planning of training activities; design and programming of training, and training process assessment) and hygiene and safety at work

(84%). Practices that apply without the involvement of HR managers (percentage above 30%) are, essentially, more administrative practices such as job planning and functional flexibility and versatility that are handled by managers or the legal department; management of salaries, bonuses and incentives for productivity and cost PM that are, in most cases, developed by the administrative department; and staff planning and functional flexibility and versatility are managed either by the manager or by middle management. Practices that do not apply to companies (percentage above 30%) are career plans, plans for downsizing, management by objectives and empowerment.

8.6.7. *Recruitment and selection*

According to the literature, recruitment and selection practices are one of the most common in SMEs. Planning the recruitment is a critical step in the strategy of companies. Poor recruitment or lack of planning (no need to fill vacancy and lack of HR) can cause various operational and financial losses for the company. Financial losses because there was an investment in a collaborator that is not going to be a surplus value to company, and operational losses because the company may require a labor force to deal with an order and did not anticipate the need for recruitment to successfully fulfill deadlines. According to Gomez-Mejia *et al.* [GOM 95], HR planning is the process that an organization uses to ensure having the right number and the right group of people to achieve a particular level of production and service in the future. Lack of HR planning by the company may cause deep damage to them. Only 34% of the respondents reported having a recruitment plan. Of these, only 4% have a plan lasting more than 5 years and 2% have a plan lasting 4–5 years. Most respondents have a short-term plan (less than 1 year for 12% and between 1 and 3 years for 16% of the respondents). One of the characteristics pointed out to HRM practices in SMEs is that these are blown business strategies, with an *ad hoc* and informal perspective (e.g. [CAR 04, BAR 07]). Results demonstrate these characteristics because most companies do not have any HR planning, working with *ad hoc* business needs.

In recruitment, selection plays an important role in finding the best employees. Among the selection methods proposed as answer options most widely used in SMEs, there are individual interviews and trial periods for all groups of employees, but with slightly different values. Selection methods used less in SMEs are handwriting analysis, tests of integrity/honesty and tests on drug use.

8.6.8. *Training and development*

Over the years, the Portuguese government has injected European funds to increase the education of the Portuguese population as well as to increase the competitiveness of companies. However, over the years, there have been criticisms about the misapplication of such funds, often unstructured, without strategic perspective and improvement of personal or organizational. Many of the millions that were spent were injected on training in SMEs. The question raised to companies was what kinds of method were used by them in training their employees. The training methods most widely used in business training were workplace (54%) and classroom training (54%). Training in the workplace can lead to huge losses for companies because they will be investing in an employee who does not fit in this kind of function or is unable to transfer the know-how to the other employees of the function. In turn, methods less used correspond to training through rotation (16%) and self-taught training (12%). These two methods seem to gain importance in future application in business (self-taught education and training through rotation, both with 16%).

According to the literature, studies on training and development in SMEs are scarce. Statistical information on training in SMEs is limited and basic research is a set an small-scale and fragmented studies [WON 97]. SMEs, due to their structure and resources, develop few training programs. It becomes difficult for company to invest in training employees, therefore does not have the HR to continue working and also because employees with low qualifications are very difficult to convince of the importance of training. However, attitudes tend to change. The support from the State Union and EU were a major contributor to the increase in training programs. Increasingly, there is training in SMEs. Despite the support given, only half of companies will moderately increase training programs. A total of 36% of companies regard it would not increase the budget for training. Portuguese entrepreneurs remain skeptical about the importance of training. Training is an important means of retaining and motivating employees.

8.6.9. *Performance*

Performance assessment is, as a rule, held annually. This assessment has several purposes, among which is assessor's opportunity to establish and revise the overall objectives and individual performance that may be maladjusted to reality of the company. According to Sousa *et al.* [SOU 06], there are several methods for performance evaluation, such as self-assessment, performed by subordinates, 360-degree assessment, etc. The most commonly used method in assessment of top managements is self-assessment (34%) and the least used method is 360-degree assessment (6%). The most used method in assessment of heads of department is assessment by superior (54%) and least used method is peer review (10%). While the assessment of personnel supervisory is held by hierarchical superior (66%), the less used method, here, is 360-degree assessment. The most used method in the assessment of operating personnel is assessment by hierarchical superior (82%), while the least used is 360-degree assessment (4%). A recent method that allows the employee to be evaluated by hierarchical superiors, himself, peers, subordinates and customers is 360-degree assessment. Its implementation is a complex method and, with regard to Sousa *et al.* [SOU 06], implementing this method of assessment in traditional organizations with high levels of mistrust and poor communication can lead to a challenge as there may be danger of failing to exploit the potential of 360-degree assessment.

8.6.10. *Incentives*

An incentive program may have several objectives: to help build and strengthen a company's culture, attract and retain the best employees, increase motivation and productivity, and strengthen the role and status of hierarchical levels [SOU 06]. Wagar and Langrock [WAG 04] suggest that the best way to predict the presence of specific compensation is the size of business, having large companies resources to invest in these programs. Most companies provide cash bonuses to their employees (32%). A total of 10% of the companies have no incentive programs. For the future, companies will invest more in insurance package and healthcare as a reward or incentive. There are no major discrepancies when analyzing practices that are currently performed and will be used in the future. At this point, 38% of the companies provide bonuses in cash and 26% in insurance packages and medical assistance. Distributions of shares have no relevance as an incentive for SMEs. Asked about what will be the trend in coming years to reward performance-based compensation, as a ratio, 68% said it will increase, and of these, 20% said it will increase significantly. A total of 26% of the respondents stated that reward based on performance would remain.**

8.7. Characterization proposal of HRM in Portuguese SMEs

This study sought to understand what is important for HRM in SMEs and realizes the existing degree of formalization, to identify a set of elements that allow defining the reality in SMEs and understanding the strengths and limitations of people management in small businesses.

By analyzing empirical results, it was possible to define the model that characterizes HRM in SMEs in Portugal (Figure 8.1).

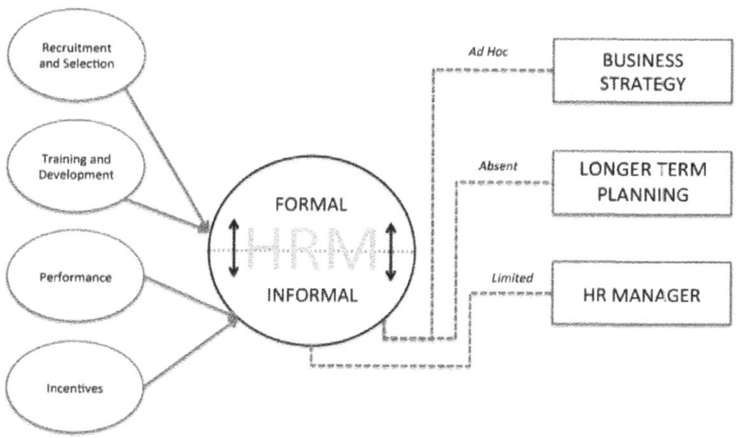

Figure 8.1. *Characterization model of HRM in SMEs in Portugal*

The model focuses on two dimensions, formal versus informal, that shapes both HRM practices and policies. Regarding HRM practices, they differ by degree of (in) formality. For example, recruitment and selection are practices with a greater degree of formality that provide incentives and rewards. On policies, we can say that these are presented informally constituted, characterized by three words "*ad hoc*", "absent" and "limited".

8.7.1. *HRM practices*

Some HRM practices are used more than others, such as recruitment and selection. Clearly, these practices have few implications in terms of resources, which facilitates their implementation.

8.7.2. *Recruitment and selection*

Recruitment and selection is, together with training and development, the most used practice in SMEs. Despite this fact, it presents a reduced formalism with a limited relationship with business strategy. For example, tools used by recruitment companies seem to be well defined for employees operating level (ad placement: 68%; personal contacts: 76%; spontaneous contacts: 80%), while for top managers this does not happen (ad placement: 22%; employment agencies: 10%; personal contacts: 30%). In SMEs, top management is very limited, consisting mainly of the administrator and company's partners. Often, for reasons of family involvement, recruitment of top managers is perpetuated by the family's company administrator.

8.7.3. *Training and development*

Training and development is the practice, together with recruitment and selection, most commonly used by SMEs. Empirical study has proven this perspective, as 82% of the companies have an effective training plan. If we cross the variables "organization has an effective training plan" with "methodology training", it shows that not only do these companies do not have a plan for ongoing training they also do not have a defined methodology. Through the intersection of the variables "the organization has an effective training plan" and "it uses professional external", we can see that it denotes that most companies resort to external training companies. Training in SMEs is described as informal and developed in the workplace, with little or no involvement of management [KOT 04]. Frequency analysis shows that 54% of the companies actually use training developed in the workplace and 38% use it, also, currently and in the future. With regard to available training, data show that training methods are diversified and structured as classroom training (54% currently use), computer-aided training (30%), with use of training videos (24%), autodidactic training (12%), personal training (40%) and training abroad (40%). Results show a relative formalization of practice, largely due to the use of specialized companies.

8.7.4. *Performance assessment*

Literature states that the application process for performance assessment in SMEs is not peaceful and there are several barriers to its development. According to Hudson *et al.* [HUD 01], the main barriers for not implementing the process of performance assessment in SMEs are limited resources and a more dynamic style and featuring in these emerging companies. These aspects are problematic because the development of process of performance assessment is necessarily long term and requires results to be strategic. The analysis of data shows that there is an association between the variable "performance assessment" and "headcount" (R Pearson from –0.067 to a significance level of 0.64). Regarding methods used by companies for the performance assessment of its employees, there is a greater focus on the use of self-assessment in top management, this group having the highest percentage of non-use of assessment. There is a higher incidence of assessment by the hierarchical superior, mainly for operational staff (82%). Methods that are more structured and require more material and financial resources, such as 360-degree assessment and assessment by customers, have a low utilization rate in all groups of employees. The results show a low formalization of performance assessment in SMEs.

8.7.5. *Incentives*

The incentives system has an important role in business. For small businesses, incentives are an important practice [CAS 02] because they often influence the efforts of SMEs in the recruitment and retention of HR [CAR 04]. The use of incentive programs, such as retention strategy, is seen by 70% of respondents as having a high or very high efficiency, with 32% of them focusing on monetary strategies. Companies that claim to have no incentive system (10%) describe retention strategies (larger benefits, pay increase, increasing number of non-monetary rewards very effective). Incentives can be monetary and non-monetary. Most companies bet more on monetary incentives, however it has acquired a non-monetary relevance. Companies are betting more on cash bonuses at the expense of non-monetary rewards. Currently, small firms are betting on bonuses in cash because they allow us, not only, to simplify the system of incentives, but also to reduce HR, here by a control made by the administrator or manager of the company. These data allow us to conclude that incentives are simple [GHO 96] and informal [WAG 04].

8.7.6. *HR policies*

HR policies look to promote the ongoing appropriateness of employee activity to achieve strategic goals and business enterprises, ensuring their professional development and valorization, as well as enhancing their technical and management expertise.

8.7.7. *Relationship with business strategy*

The new interest in HR as strategic to the organization seeks to fit HR with an important role in the creation of wealth, directly contributing to the implementation of strategic and operational objectives for the organization

[BEC 96]. Based on the variables "overall business strategy" and "HR manager's role", we can see that respondents who define HRM as totally related to overall business strategy identify the HR manager as a business partner. Asked about what are the most critical aspects of HRM for the near future, respondents focus on customer satisfaction and improved employee productivity. To investigate this issue, we have crossed the variables "performance assessment" and "productivity improvement". A total of 65% of the companies present, as most critical to improving their productivity, the need to develop a performance assessment program, with the involvement of the HR department. The same happens with the intersection of the variables "productivity bonuses" and "performance bonuses" with "productivity improvements". A total of 59% of the companies implement productivity bonuses with the involvement of the HR department and 58% implement performance bonuses. Results are similar when we use the variables "employee satisfaction" and "importance of people". A total of 44% of the companies that define employee satisfaction as critical to company claim to have employees as very important for long-term planning. One of the characteristics of HRM practices in SMEs is that these practices are absent from business strategies, standing out from an *ad hoc* and informal perspective (e.g. [CAR 04, BAR 07]). These results confirm these characteristics, because most companies do not have any HR planning, working from *ad hoc* business needs.

8.7.8. *Long-term planning*

In order to analyze the HR role in long-term planning, a Friedman test was carried out. The results, statistically significant at a significance level of 0.01, confirm that there is a clear preference for technology variables (3.96) and capital (4.96) instead of people (5.04).

8.7.9. *HR manager role*

Through the intersection of the variables "general business strategy" and "HR manager's role", we can see that respondents who define HRM as totally related to overall business strategy identify the HR manager as a business partner. More than 50% of the respondents have this scenario as a future in the company. Respondents describing the association between HRM and general business strategy as being more or less related, identifying the HR manager as an administrative specialist and a change agent, also present this as a future aspiration.

8.8. Conclusions

The present study aimed to characterize HRM in SMEs in Portugal, through the analysis of HRM policies and practices used and the degree of formalization. This research was relevant because knowledge of the reality of HRM in SMEs in Portugal is non-existent. Even at an international level, research is scarce when directed to HRM in SMEs. There is a set of fragmented studies, which do not allow isolating accurate information about the reality of SMEs. Studies by Cardon and Stevens [CAR 04] and Cassel *et al.* [CAS 02] allowed another view on the impact of HRM practices on the reality of SMEs to be given. Cardon and Stevens [CAR 04] report that all SMEs have some form of HRM, even if informal. In addition, according to Cassel *et al.* [CAS 02] there is a great diversity in the implementation of HRM practices in SMEs and their relative success. An SME's approach to HRM is performed case by case and reactively, as opposed to a holistic, proactive or systemic.

The model that we propose as a possible characterization of Portuguese reality presents a constant image of informality of procedures pertaining to HRM practices. Only recruitment and selection and training and development, which present some evidence, have formalized procedures. Still, we cannot say that they present strategic outcomes for business. Both relationships with business strategy as well as long-term planning are almost non-existent. This is reflected in the importance given to the HR manager, who frequently appears with a small strategic and administrative dimension.

8.9. Bibliography

[ARM 00] ARMSTRONG M., "The name has changed but the games remained the same?", *Employee Relations*, vol. 22, no. 6, pp. 576–593, 2000.

[AUD 01] AUDRETSCH D.B., THURIK A.R., "What is new about the new economy: sources of growth in the managed to the entrepreneurial economy", *Journal of Evolutionary Economics*, vol. 10, no. 1, pp. 17–34, 2001.

[BAR 99] BARBER A.E., WESSON M.J., ROBERSON Q.M., *et al.*, "A tale of two job markets: organizational size and its effects on hiring practices and job search behavior", *Personnel Psychology*, vol. 52, pp. 841–867, 1999.

[BAR 98] BARNEY J.B., WRIGHT P.M., "On becoming a strategic partner: the role of human resources in gaining competitive advantage", *Human Resource Management Journal*, vol. 37, pp. 31–46, 1998.

[BAR 03] BARON R.A., "Human resource management and entrepreneurship: some reciprocal benefits of closer links", *Human Resource Management Review*, vol. 13, pp. 253–256, 2003.

[BAR 07] BARRETT R., MAYSON S., "Human resource management in growing small firms", *Journal of Business and Enterprise Development*, vol. 14, no. 2, pp. 307–320, 2007.

[BEC 96] BECKER B., GERHART B., "The impact of human resource management on organizational performance: progress and prospects", *Academy of Management Journal*, vol. 39, no. 4, pp. 779–801, 1996.

[BRE 93] BREWSTER C., "Developing a "European" model of human resource management", *The International Journal of Human Resource Management*, vol. 4, no. 4, pp. 765–784, 1993.

[BRE 04] BREWSTER C., "European perspectives on human resource management", *Human Resource Management Review*, vol. 14, pp. 365–382, 2004.

[BRE 07] BREWSTER C., "Comparative HRM: European views and perspectives", *International Journal of Human Resource Management*, vol. 18, no. 5, pp. 769–787, 2007.

[CAB 99] CABRAL-CARDOSO C., "Gestão de Recursos Humanos: Evolução do conceito, perspectivas e novos desafios", in CUNHA M. (ed.), *Teoria Organizacional*, Publicações Dom Quixote, Lisbon, pp. 225–249, 1999.

[CAB 04] CABRAL-CARDOSO C., "The envolving Portuguese model of HRM", *International Journal of Human Resource Management*, vol. 15, no. 6, pp. 959–977, 2004.

[CAB 06] CABRAL-CARDOSO C., "Portuguese management between global rhetoric and local reality", *Management Research*, vol. 4, no. 3, pp. 193–204, 2006.

[CAE 07] CAETANO A., VALA J., *Gestão de recursos humanos: contextos, processos e técnicas*, Editora RH, Lisbon, 2007.

[CAR 04] CARDON S.M., STEVENS C.E., "Managing human resources in small organizations: what do we know?" *Human Resource Management Review*, vol. 14, pp. 295–323, 2004.

[CAS 02] CASSEL C., NADIN S., GRAY M., *et al.*, "Exploring human resource management practices in small and medium sized enterprises", *Personnel Review*, vol. 31, no. 6, pp. 671–692, 2002.

[CHU 83] CHURCHILL N., LEWIS V., "The five stages of small business growth", *Harvard Business Review*, vol. 61, no. 3, pp. 30–50, 1983.

[CUN 05] CUNHA M.P., "Adopting or adapting? The tension between local and international mindsets in Portuguese management", *Journal of World Business*, vol. 40, pp. 188–202, 2005.

[CUN 07] CUNNINGHAM L.X., ROWLEY C., "Human resource management in Chinese small and medium enterprises: a review and research agenda", *Personnel Review*, vol. 36, no. 3, pp. 415–439, 2007.

[DEN 03] DENNIS W.J., "Raising response rates in mail surveys of small business owners: results of an experiment", *Journal of Small Business Management*, vol. 41, no. 3, pp. 278–295, 2003.

[DUA 02] DUARTE N., "Criação de PMEs na região Norte no âmbito do programa RIME durante o QCA II", *Gestão e Desenvolvimento*, vol. 11, pp. 57–80, 2002.

[DUB 95] DUBERLEY J.P., WALLEY P., "Assessing the adoption of HRM by small and medium-sized manufacturing organizations", *International Journal of Human Resource Management*, vol. 4, no. 4, pp. 891–909, 1995.

[FRA 98] FRAZA V., "No small feat", *Industrial Distribution*, vol. 87, pp. 48–49, 1998.

[GHO 96] GHOBADIAN A., GALLEAR D.N., "Total quality management in SMEs", *International Journal of Management Science*, vol. 24, no. 1, pp. 83–106, 1996.

[GOM 95] GÓMEZ-MEJIA L.R., BALKIN D.B., CARDY R.L., *Managing Human Resources*, Prentice-Hall, Englewood Cliffs, NJ, 1995.

[GRE 72] GREINER L., "Evolution and revolution as organizations grow", *Harvard Business Review*, vol. 50, no. 4, pp. 37–46, 1972.

[GUE 87] GUEST D.E., "Human resource management and industrial relations", *Journal of Management Studies*, vol. 24, no. 5, pp. 503–521, 1987.

[GUE 89] GUEST D.E., "Personnel and HRM: can you tell the difference?", *Personnel Management*, vol. 21, no. 1, pp. 48–51, 1989.

[GUE 90] GUEST D.E., "Human resource management and the American dream", *Journal of Management Studies*, vol. 27, no. 4, pp. 377–397, 1990.

[HEN 86] HENDRY C., PETTIGREW A., "The practice of strategic human resources management", *Personnel Review*, vol. 15, pp. 2–8, 1986.

[HEN 99] HENEMAN H.G., BERKLEY R.A., "Applicant attraction practices and outcomes among small businesses", *Journal of Small Business Management*, vol. 37, no. 1, pp. 53–74, 1999.

[HOL 84] HOLLAND P.G., BOULTON W.R., "Balancing the family and the business in family business", *Business Horizons*, vol. 27, pp. 16–21, 1984.

[HUD 01] HUDSON M., SMART A., BOURNE M., "Theory and practice in SME performance measurement systems", *International Journal of Operations & Production Management*, vol. 21, no. 8, pp. 1096–1115, 2001.

[HUS 95] HUSELID M. "The impact of human resource management practices on turnover, productivity, and corporate financial performance", *Academy of Management Journal*, vol. 38, no. 3, pp. 635–672, 1995.

[IAP 05] IAPMEI, Nova definição de PME: Recomendação 2003/361/CE, 2005. Available at http://www.eicpme.iapmei.pt/index.php (accessed on 12 June 2012).

[JAC 89] JACKSON S.E., SCHULER R.S., RIVERO J.C., "Organizational characteristics as predictors of personnel practices", *Personnel Psychology*, vol. 42, pp. 727–786, 1989.

[KOK 01] KOK J., UHLANER L.M., "Organization context and human resource management in the small firm", *Small Business Economics*, vol. 17, pp. 273–291, 2001.

[KOT 04] KOTEY B., SHERIDAN A., "Changing HRM practices with firm growth", *Journal of Business and Enterprise Development*, vol. 11, no. 4, pp. 474–485, 2004.

[MAC 95] MACDUFFIE J.P., "Human resources bundles and manufacturing performance: organizational logic and flexible production systems in the world auto industry", *Industrial and Labor Relations Review*, vol. 48, no. 2, pp. 197–221, 1995.

[MCE 84] MCEVOY G.M., "Small business personnel practices", *Journal of Small Business Management*, vol. 22, pp. 1–8, 1984.

[MCE 00] MCELWEE G., WARREN L., "The relationship between total quality management and human resource management in small and medium-sized enterprises", *Strategic Change*, vol. 9, no. 7, pp. 428–435, 2000.

[NIK 05] Nikandrou I., Apospori E., Papalexandris N., "Changes in HRM in Europe: a longitudinal comparative study among 18 European countries", *Journal of European Industrial Training*, vol. 29, no. 7, pp. 541–560, 2005.

[OEC 04] Oecd, "Promoting entrepreneurship and innovative SMEs in a global economy: towards a more responsible and inclusive globalisation", *2nd OECD Conference of Ministers Responsible for Small and Medium-Sized Enterprises*, Istanbul, 2004.

[RAM 91] Ram M., "Control and autonomy in small firms: the case of the West Midlands clothing industry", *Work Employment and Society*, vol. 5, no. 4, pp. 601–619, 1991.

[RAM 93] Ram M., Holliday R., "Relative merits: family culture and kinship in small firms", *Sociology*, vol. 27, no. 4, pp. 629–648, 1993.

[REI 01] Reid R.S., Adams J.S., "Human resource management – a survey of practices within family and non-family firms", *Journal of European Industrial Training*, vol. 25, no. 6, pp. 310–320, 2001.

[ROT 98] Roth P.L., Bevier C.A., "Response rates in HRM/OB survey research: norms and correlates, 1990–1994", *Journal of Management*, vol. 24, no. 1, pp. 97–117, 1998.

[SOU 06] Sousa M.J., Duarte T., Sanches P.G., *et al.*, *Gestão de recursos humanos: métodos e práticas*, Lidel, Lisbon, 2006.

[STA 05] Stavrou-costea E., "The challenges of human resource management towards organizational effectiveness: a comparative study in Southern EU", *Journal of European Industrial Training*, vol. 29, no. 2/3, pp. 112–182, 2005.

[STO 94] Storey D.J., *Understanding the Small Business Sector*, Routledge, London, 1994.

[TOR 89] Torrington D., "Human resource management and the personnel function", in Storey J. (ed.), *New Perspectives on Human Resource Management*, Routledge, London, 1989.

[UNE 06] Unep, Big challenge for small business: sustainability and SMEs. Industry and Environment, vol. 26, 2006. Available at http://www.uneptie.org/media/review/vol26no4/IE26_4-SMEs.pdf (accessed on 27 June 2012).

[WAG 98] Wagar T., "Determinants of human resource management in small firms: some evidence from Atlantic Canada", *Journal of Small Business Management*, vol. 36, pp. 13–23, 1998.

[WAG 04] Wagar T., Langrock L., "Performance appraisal and compensation in small firms", *Canadian HR Reporter*, vol. 12, p. 10, 2004.

[WIL 99] Wilkinson A., "Employment relations in SMEs", *Employee Relations*, vol. 12, no. 3, pp. 206–217, 1999.

[WON 97] Wong C., Marshall J.N., Aldermart N., *et al.*, "Management training in small and medium-sized enterprises: methodological and conceptual issues", *The International Journal of Human Resource Management*, vol. 8, no. 1, pp. 44–65, 1997.

List of Authors

M. Reza ABDI
School of Management
Bradford University
West Yorkshire
United Kingdom

Alex Lôbo CARLOS
Brasil Navy
and
Department of Management
School of Economics and Management
University of Minho
Braga
Portugal

J. Paulo DAVIM
Department of Mechanical Engineering
University of Aveiro
Portugal

Regina Maria DE OLIVEIRA LEITE
Department of Management
School of Economics and Management
University of Minho
Braga
Portugal

Frank E.P. DIEVERNICH
Lucerne University of Applied Sciences and Arts
Bern
Switzerland

João FONTES DA COSTA
Oliveira do Hospital High School of Technology and Management
Polytechnic Institute of Coimbra
Oliveira do Hospital
Portugal

Stuart HOLLAND
Faculty of Economics
University of Coimbra
Portugal

Vipin KHANNA
School of Management
Bradford University
West Yorkshire
United Kingdom

Carolina Feliciana MACHADO
Department of Management
School of Economics and Management
University of Minho
Braga
Portugal

Pedro Ribeiro NOVO DE MELO
Department of Management
School of Economics and Management
University of Minho
Braga
Portugal

Wayne O'DONOHUE
Department of Employment Relations and Human Resources
Griffith Business School
Griffith University
Queensland
Australia

Maria João SANTOS
Higher Institute of Economics and Management
Lisbon Technical University
Portugal

Kim Oliver TOKARSKI
Faculty of Business
Bern University of Applied Sciences
Bern
Switzerland

Nuttaneeya (Ann) TORUGSA
Australian Innovation Research Centre
University of Tasmania
Australia

Teresa Carla TRIGO OLIVEIRA
Faculty of Economics
University of Coimbra
Portugal

Raky WANE
SOCIUS Research Center
Higher Institute of Economics and Management
Lisbon Technical University
Portugal

Index

Printed and bound by CPI Group (UK) Ltd, Croydon, CR0 4YY

13/09/2026

14961076-0004